KNOWLEDGE-BASED REACTIVE SCHEDULING

IFIP Transactions B: Applications in Technology

International Federation for Information Processing

Technical Committee 5:

Computer Applications in Technology

IFIP Transactions Abstracted/Indexed in:
INSPEC Information Services
Index to Scientific & Technical Proceedings®
CompuMath Citation Index®, Research Alert™, and SciSearch®

B-15

KNOWLEDGE-BASED REACTIVE SCHEDULING

Proceedings of the IFIP TC5/WG5.7 International Workshop on
Knowledge-Based Reactive Scheduling
Athens, Greece, 1 October, 1993

Edited by

ELIZABETH SZELKE

Computer and Automation Research Institute
Hungarian Academy of Sciences
Budapest, Hungary

ROGER M. KERR

School of Mechanical and Manufacturing Engineering
University of New South Wales
Kensington, NSW, Australia

1994

NORTH-HOLLAND
AMSTERDAM • LONDON • NEW YORK • TOKYO

ELSEVIER SCIENCE B.V.
Sara Burgerhartstraat 25
P.O. Box 211, 1000 AE Amsterdam, The Netherlands

Keywords are chosen from the ACM Computing Reviews Classification System, ©1991, with permission.
Details of the full classification system are available from
ACM, 11 West 42nd St., New York, NY 10036, USA.

ISBN: 0 444 81814 6
ISSN: 0926-5481

This book is printed on acid-free paper.

Printed in The Netherlands

FOREWORD

The problem of reactive scheduling, although very familiar to those actually responsible for executing schedules, has up to now received relatively little explicit attention in the literature. The complementary problem of constructing *good predictive* schedules for operational activities in the presence of many complex and often conflicting constraints, has on the other hand been a focus of research attention for a number of years. We are now however beginning to see a gradual change in this situation, and an increasing number of journal articles and conference papers are starting to concern themselves specifically with reactive scheduling, its special problems, and its relationship to predictive scheduling.

Reactive scheduling is concerned with monitoring and controlling the execution of predictive schedules and making any changes required to bring them in line with unanticipated events or disturbances happening in real time. It is an activity that is closely allied to predictive scheduling but it has the added dimension of stringent real time execution requirements that constrain the nature and extent of any computations that can occur. More than ever, in reactive scheduling, we are required to pay attention to the computational efficiency of the algorithms or heuristics invoked to choose the best response to a real time event. Allied to this is the need to develop predictive schedules that are robust against a wide range of contingencies and to investigate trade-offs between robustness and cost.

The purpose of this workshop is to explore the scope and status of both existing and potential approaches to reactive scheduling. Although the term "knowledge-based" has become somewhat of a buzz-word, the relevance and importance to reactive scheduling of domain knowledge, whether stored in computers or in the brains of experienced schedulers, is inescapable, and so the workshop has been deliberately focussed on knowledge based approaches to the problem. The workshop contributions examine a variety of approaches ranging from reactive scheduling decision support frameworks in which the knowledge comes mainly from the human through to completely new approaches to automated reactive scheduling such as neural networks whose potential in this field is as yet relatively unexplored. It is to be hoped that the workshop will stimulate an increased amount of discussion on the important and emerging topic of knowledge based reactive scheduling.

Roger M. Kerr
Workshop Chairman
School of Mechanical and Manufacturing Eng.
University of New South Wales
Kensington, Australia

Elizabeth Szelke
Workshop Coordinator
Computer and Automation Res. Inst.
Hungarian Academy of Sciences
Budapest, Hungary

CONTENTS

Knowledge-Based Reactive Scheduling (B-15)
E. Szelke and R.M. Kerr (Editors)
Elsevier Science B.V. (North-Holland)

Knowledge-based Reactive Scheduling in a Flexible Manufacturing System

Klaus Fischer

DFKI, Project AKA-MOD, Stuhlsatzenhausweg 3, 66123 Saarbrücken,
e-mail: kuf@dfki.uni-db.de

Abstract

This paper presents a framework for the design of the planning and controlling components in a flexible manufacturing system. A hierarchical planning structure is presented which consists of six layers: the layer of the production planning and control system, the layer of the shop floor control system, the task coordination layer, the task planning layer, the task execution layer and the machine control layer. The last four layers mentioned are present separately for each manufacturing device which is able to execute tasks autonomously. For the abstract system which consists of the task coordination component, the task planner, the task execution component, the machine control component, and the flexible manufacturing device itself the term *autonomous system* is introduced. Special emphasis is placed on solving the scheduling problem in the shop floor control system and task planning in the task planning layer of the autonomous systems. The implementation of the planning and controlling components of the hierarchical planning structure was done for the example of the model of a manufacturing plant [4, 7] with the help of the rule-based multi-agent system MAGSY [8, 6], which is briefly described at the end of the paper.

Keywords: knowledge-based, reactive, scheduling, production planning, production control, computer-aided manufacturing, flexible manufacturing system, autonomous systems, task planning, multi-agent systems, rule-based programming.

1. Introduction

CIM (Computer-Integrated Manufacturing) is nowadays a catchword to which current research work is trying to give real contents [20]. In the environment of a laboratory it is already possible to carry out a manufacturing order fully automatically, from the layout of the product to the manufacturing of the end product. At the beginning of the research work, success in special subjects, e.g. CAD (Computer-Aided Design) systems, had been achieved. The main research work today is on the integration of the units of a flexible manufacturing system (FMS). At first the main problem was the communication between these units. The ongoing connection of these units in local area networks and the definition of communication protocols has brought about great progress here. Now it is possible to concentrate on the real problems in the planning and controlling components

Figure 1. Picture of the model of a manufacturing plant.

of a FMS. These are the control of the flow of information and the control of the movement of the workpieces and tools in the manufacturing floor. In order to control the movement of the workpieces and tools, it must be determined how and when they have to be carried from one location to another. Controlling the flow of information requires ensuring that the data guiding the manufacturing process, e.g. how a workpiece has to be worked on, accompanies with the physical transport of the workpieces. The concept presented in this paper is a framework for the design of the planning and controlling components in a FMS.

The implementation of the planning and controlling components of the hierarchical planning structure which is outlined in this paper was done for the example of the model of a manufacturing plant [4, 7] with the help of the rule-based multi-agent system MAGSY [8, 6], which is briefly described at the end of the paper. This model (of the size of 140x120 cm^2) was built with fischertechnik bricks in the laboratory of the Institute for Real-time Systems and Robotics at the Technische Universität in Munich (see figure 1). Two robots, each of which has four axes, serve machine tools when performing pick-and-place tasks. Depot_1 can be handled only by robot_1; on the other side, the three machine tools can only be served by robot_2. The working spaces of the two robots overlap in the area of the input conveyor belt, the depot_2, the heating cell and the output conveyor belt. Therefore, it is necessary to give the robots strategies to avoid collision. When serving the machine tools and loading the heating cell, the movements of the robots have to be coordinated with the activities of the machines. The workpieces enter the manufacturing plant over the input conveyor belt, where they are identified by a barcode decoder. With the help of inductive sensors, which are attached to the conveyor belts as well as to the workpiece spots of the machines, it is possible to check the presence of a workpiece. Depot_1 takes up workpieces which cannot be worked on at the moment. Depot_2 is used to pass over workpieces from robot_1 to robot_2.

The implementation of the planning and controlling hierarchy for this model was a complex task. To have a comfortable environment to develop and test our implementation we built up a simulation environment using the universal simulation system USIS, which was developed at the Institut für Werkzeugmaschinen und Betriebswissenschaften at the Technische Universität in Munich.

2. Planning and Controlling Layers in a Flexible Manufacturing System

The complex task of planning and controlling the manufacturing process can be distributed among various layers: production planning and control, shop floor control, task coordination, task planning and machine control. Each of these layers has a clearly defined scope of competence. Splitting the act of planning and controlling into different layers brings the additional advantage that the deeper layers can be accessed directly without using the upper layers. Using the machine control layer, a machine can be prepared for a special task. On the task planning layer, a human operator may serve a machine tool with a workpiece. With manufacturing orders of high priority, it may be helpful to be able to pass them to the shop floor control layer directly, without using the production planning control layer. It is only important that the actions which access the deeper layers directly do not hamper the correct execution of the tasks in the upper layers. In the literature [10, 12, 16] there is agreement only on the specification of the upper two layers. The definition of the lower layers varies, and it is often restricted to special hardware equipment. In the concept presented by this paper, the definition of the lower layers is based on the robots and machines being able to plan and execute their tasks autonomously.

2.1. The Production Planning and Control Layer

The production planning and control (PPC) system is the highest controlling body to reach the goals defined by the management of a company. The registration of client orders takes place in the PPC system where they are treated by economical criteria. Here the classical functions: *administration of client orders, data management, material management*, and *time management*, are fulfilled. The horizon of planning is a day up to a week. Planning of a specific client order is done with respect to the finishing date and the sequence of the working steps to be done. In doing so a rough estimation of the capacities of the resources needed for this client order is computed [16]. By an off-line planning step working plans which determine the manufacturing sequence for a certain end product are given to the PPC system. These working plans are passed as manufacturing orders to the shop floor control system where they are executed. All economical aspects in planning are done by the PPC system and during the design of the working steps and working plans by a product engineer. For the deeper planning layers it is above all important to execute the manufacturing orders defined by the PPC system efficiently with respect to the time constraints specified by the PPC system. The more deterministic the execution of the manufacturing orders takes place, the better it is possible for the PPC system to plan the

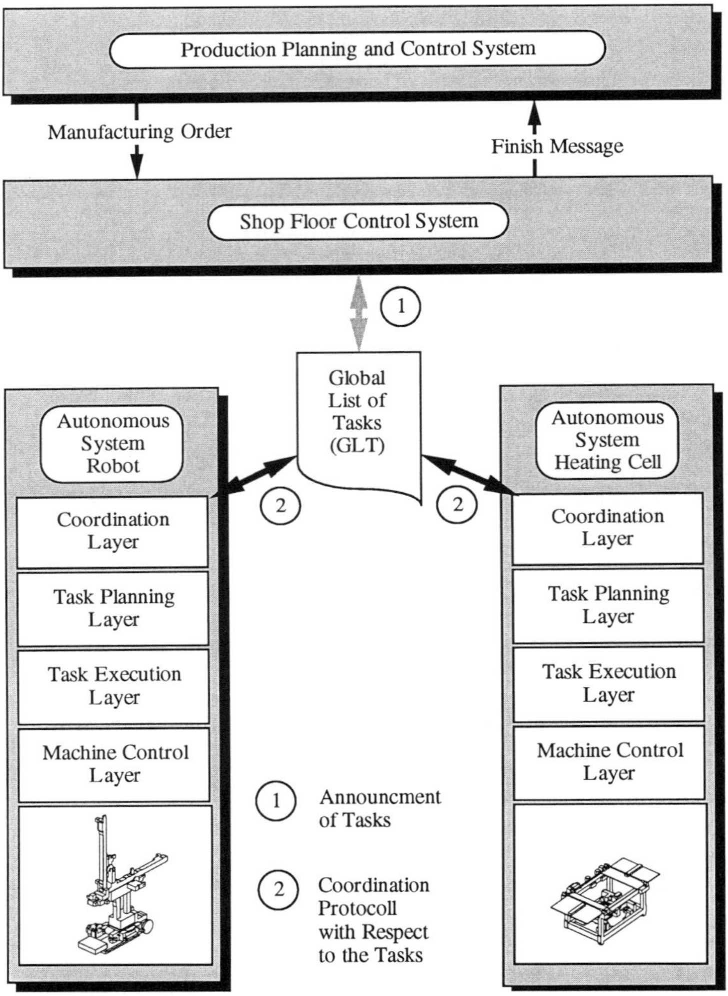

Figure 2. A hierarchical planning structure for a flexible manufacturing system.

execution of the client orders. Because planning is done using statistical data, variations in the execution of the manufacturing orders due to autonomous planning decisions can be taken into account.

2.2. The Shop Floor Control Layer

At a certain point of time, a set of manufacturing orders is present in the shop floor control (SFC) system. These manufacturing orders have to be executed in consideration of the time constraints specified by the PPC system. For each manufacturing order the due date is specified by the PPC system. The release date to start a manufacturing order is determined by the moment at which the manufacturing order is passed from the PPC system to the SFC system. Because all the economical aspects of a manufacturing order are considered in the PPC system, the main problem to be solved in the SFC system is how to execute the manufacturing orders efficiently and how to meet the time constraints specified by the PPC system. There are in general two approaches to solve this problem. In the first one a schedule for a period of time (e.g. a week or a month) is computed off-line. The schedule is then executed when the planning period passes by. In the second approach an overall schedule for longer period in time is not computed. Instead, the scheduling is done on-line in a reactive manner. Considerable work on the first approach has already been done (e.g. [9, 19, 21]) whereas only little work has been done on the second approach.

The central problem to be solved in the SFC system is the control of the flow of material. There are two conflicting goals to be reached in controlling a flexible manufacturing system: *minimal costs of inventory by minimizing the duration of the manufacturing orders in the shop floor* and *maximal rate of capacity utilization of the machines.* It is impossible to satisfy both of these two goals equally good [17]. Therefore, a compromise has to be found between these two goals. By increasing the technological level, the resource time becomes more and more important for competitiveness because advantages in time can become advantages in costs. Hence, saving of time is an important aspect in a flexible manufacturing system and, therefore, it is a common agreement that a good strategy is to try to reach the first of the two goals and to get additionally a reasonable good capacity utilization of the machines. The global goal here is to have a high flexibility in the design of products and in the manufacturing process of these products, and thus, to be able to react quickly to the changes of the market.

By controlling the flow of material the main problem to be solved is the management of the resources and their allotment to the different manufacturing orders. Here, a set of manufacturing orders

$$A_1, \ldots, A_n, n \in I\!N$$

is given. For each manufacturing order a sequence of working steps

$$A_1 = (S_{1,1}, \ \ldots, S_{1,m_1}), \ \ldots, A_n = (S_{1,1}, \ \ldots, S_{1,m_n}), \ m_1, \ldots, m_n \in I\!N$$

is given, which have to be executed on a set of machines $M_1, \ldots, M_k, k \in I\!N$. The problem to distribute this set of manufacturing orders to the set of machines is already for $k = 2$ NP hard [13]. In a flexible manufacturing system the problem is not only to distribute a set of manufacturing orders to a set of machines. Additionally, manipulation and transportation tasks have to be planned. This fact increases the complexity of the problem. Because the period in time a working, manipulation or transportation task will last is not known exactly, a global optimal plan suffers from high uncertainties. These uncertainties make it very unlikely that — even if it would be possible to compute an optimal plan — the produced plan is really optimal when it is executed and therefore make a big computational effort very doubtful. Because the time complexity of the problem is as bad as explained above, algorithms solving this problem normally use heuristics to get a reasonable good suboptimal plan. Our opinion is that, because of their simplicity, these heuristics can be computed on-line, parallel to the execution of the production process in the shop floor using the actual time constraints given by the production process. Therefore, we do not want to compute an optimal plan in our system. Instead we built up a reactive system which makes its planning decisions by using heuristics with respect to the actual situation in the shop floor.

$$D(A_i) = \sum_{k=1}^{m_i} E_B(S_{i,k}) + E_W(S_{i,k})$$

is a function which estimates the duration of a manufacturing order, where

$E_B(S_{i,k})$: Estimated duration of working step $S_{i,k}$

$E_W(S_{i,k})$: Estimated time step $S_{i,k}$ has to wait until it can be started

$$D_j(A_i) = \sum_{k=1}^{j} E_B(S_{i,k}) + E_W(S_{i,k})$$

is a function which estimates the time which is needed until the j-th working step in the working plan of the manufacturing order A_i is finished. It is not very complicated to determine the estimated values $E_B(S_{i,k})$ and $E_W(S_{i,k})$ for a working step $S_{i,k}$, as a simple approach we can use the average values determined in the past. When the system is started no such values form the past will be available, therefore, the system is started with pessimistic estimations of these values.

For each manufacturing order A_i the following functions are defined

Order id:	$number(A_i) = i$
Starting time of the order:	$start(A_i)$
Number of the already executed working steps:	$step(A_i)$
When should the next working step be started:	$predicted(A_i) = D_{step(A_i)}(A_i) + E_W(S_{i,step(A_i)+1})$

Let t_{cur} denote the current time, then we can state

$t_{cur} - start(A_i) < predicted(A_i) \Rightarrow$ Manufacturing order A_i is well in time and should be executed if there are resources available.

$t_{cur} - start(A_i) \geq predicted(A_i) \Rightarrow$ Manufacturing order A_i has problems with its time constraints and should be executed as fast as possible.

Let $\mathcal{W}_M$ be the set of all manufacturing orders which is not currently worked on an the next working step of which must be executed on machine M. A good criterion to select the next manufacturing order which should be executed next on machine M can be specified by

$$g(\hat{A}, \mathcal{W}_M) = \sum_{A \in \mathcal{W}_M \setminus \hat{A}} (margin(A) - E_B(S_{number(\hat{A}), step(\hat{A})+1})) \rightarrow max,$$

where

$$margin(A) = predicted(A) - (t_{cur} - start(A))$$

Is there more than one manufacturing order with the maximum value of function g, then one of these manufacturing orders may be chosen freely. If for a machine M holds:

$$\forall \hat{A} \in \mathcal{W}_M : \exists A \in \mathcal{W}_M : \hat{A} \neq A \wedge (margin(A) - E_B(S_{number(\hat{A}), step(\hat{A})+1})) < 0$$

then it is overloaded, i.e. no matter which of the manufacturing orders waiting for machine M is chosen next, one of the other manufacturing orders will run into problems with respect to its time constraints. In practice it is sufficient to check only the following condition which is easier to compute:

$$\exists A \in \mathcal{W}_M : \hat{A} \neq A \wedge (margin(A) - E_B(S_{number(\hat{A}), step(\hat{A})+1})) < 0$$

where $\hat{A}$ is manufacturing order selected by the goal function g to be executed next on machine M. If this condition is fulfilled there is at least one manufacturing order which has problems with respect to its time constraints. For this manufacturing order it is likely that the time constraints which were specified by the PPC system will be violated. This fact has to be reported to the PPC system to give it the possibility to react to it.

In addition to time constraints specified for the manufacturing order by the PPC system a system engineer should be able to specify priorities for the manufacturing orders to be able to influence the sequence of their execution. The goal function g can be modified in the following manner to take care of priority specifications:

$$g(\hat{A}, \mathcal{W}_M) = \sum_{A \in \mathcal{W}_M \setminus \hat{A}} prio(A)^{-margin(A)} (margin(A) - E_B(S_{number(\hat{A}), step(\hat{A})+1})) \rightarrow max$$

In practice it is not sufficient that in a FMS time problems in the execution of the manufacturing orders are recognized and signaled to the PPC system. It must also be possible to ask for the current distribution of the manufacturing orders in the shop floor [15]. Normally, the visualization of the current production situation is done in a Gant charts [22]. Using the strategies mentioned above, a global schedule for the distribution of the manufacturing orders to the machines is not computed explicitly. Therefore, it is only possible to make a forecast of the events which are likely to take place in the future. To be able to compute such a forecast the SFC system collects statistical data to be able to compute the expected values for the time necessary for the execution of a certain task. Using this statistical data it is possible to compute a forecast of the events

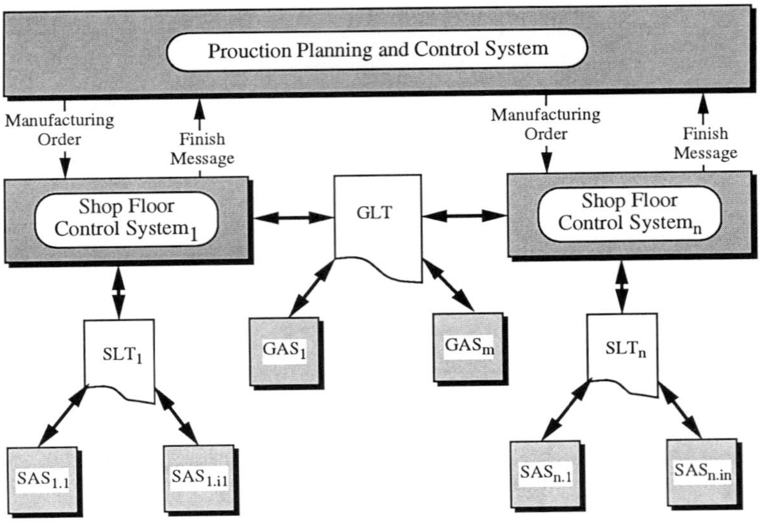

Figure 3. A shop floor control system is introduced for each production unit.

which are likely to happen in the near future. In this forecast the physical constraints of the production system have to be recognized. For example it is not possible to transport two manufacturing orders in parallel if there is only one FTS (flexible transport system) which is not able to transport both manufacturing orders at the same time.

In figure 2 all autonomous systems (AS) (see also 2.3) are subordinated to one SFC system. For small applications like the one described in the introduction this model is sufficient [14]. In practice a factory consists of several buildings which represent independent production units. In this case it is not very meaningful to control all of these production units by one SFC system. More likely a SFC system is introduced for each of these production units (see figure 3).

Even inside a production unit it may be not meaningful to subordinate all ASs immediately to the SFC system. Inside a production building there may be groups of ASs which work together closely. The members of these groups are normally not mobile. An example for such a group may be a machine tool and a robot fixed to this machine for serving it. Such a group of ASs shall be called a *flexible cell.* Normally, for a flexible cell it is possible to determine the sequence of the execution of manufacturing orders by itself. Because changes in the configuration of a machine will slow down the execution of a sequence of manufacturing orders in a flexible cell, one tries to find an execution sequence where the changes of the configuration of the ASs in a flexible cell is minimal [2, 11]. In doing so the time constraints specified by the superordinated SFC system have to be considered.

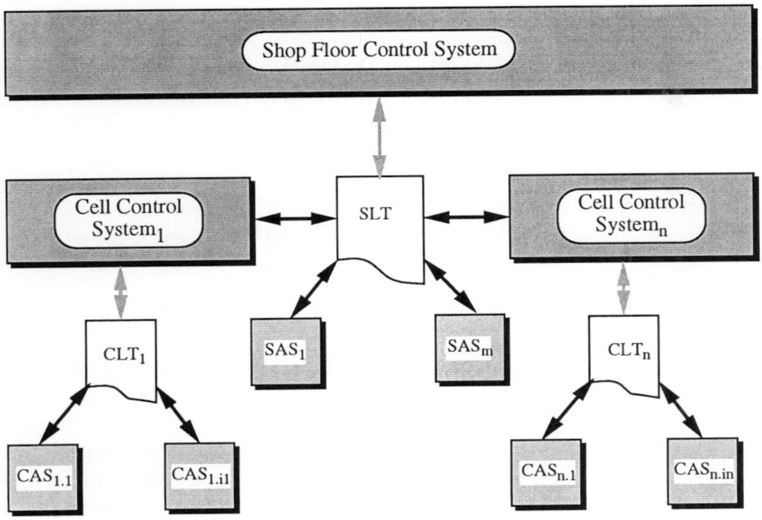

Figure 4. For groups of ASs which work together closely are cell control systems introduced.

It is not very meaningful to do the computation of an optimal execution sequence in the SFC system of a production unit itself. Generally, this problem is again NP hard [11]. This leads to the idea to introduce a control system for each flexible cell separately which is called flexible cell control (FCC) system (see figure 4).

If we structure the SFC layer in this manner, three kinds of ASs can be distinguished:

- ASs which are not related to a SFC system.
- ASs which are directly related to a SFC system.
- ASs which are related to a FCC system.

Corresponding to this there is one global list of tasks (GLT), one list of tasks for each SFC system (SLT), and one list of task for each FCC system (CLT) (see figure 3 and 4). A control system is allowed to cast tasks to the task list which is related to this control system and to to the task list of the superordinated control system. Therefore, a FCC system is able to cast tasks to its own CLT and to the SLT of the SFC system it is related to. A SFC system is able to cast tasks to its own SLT and to the GLT.

The tasks which are put to the GLT by a SFC system are normally transport tasks between to production units. Typically such a transportation task can be done by a single AS. Therefore, there is no cooperation necessary and, hence, the AS executing such a task can select this task purely by local criteria of optimality. Of course, the time constraints specified with the tasks have to be considered.

If a FCC system puts a task to the SLT of the superordinated SFC system, a group of ASs which are related to the superordinated SFC system which are able to solve this task has to be found. Only when such a group of ASs is found the cell is internally prepared to solve this task. This is done by putting the corresponding task to the related CLT. The FCC system has to be aware that for a certain task a the ASs of the cell which are able to serve this task remain not bound to the currently executed task far beyond the point in time when task a is put to the SLT of the superordinated SFC system.

2.3. The Task Coordination Layer

Looking at the hierarchical planning structure, the task coordination layer (TCL) is the first planning and controlling layer which is present for each autonomous unit separately. The planning and controlling layers from the TCL to the machine control layer (MCL) group to a logical unit which is unified with the term 'autonomous system' (AS). The SFC system passes the task to the TCL as soon as it is determined by the production plan that the task may be executed because all of the preceding working steps have been completed. The SFC system does not care if it is possible for a group of ASs to execute this task immediately or if they are currently engaged in the execution of a task. The task is just announced by the SFC system and the ASs decide by themselves when it will actually be executed. By doing a specific task several ASs have to cooperate. Each AS has to play a part to solve a specific task. No AS may believe that it is the only one which wants to play a certain part for a specific task. Therefore, the ASs must coordinate their intentions of playing parts in different tasks on the TCL. The main problem to be solved on the TCL is the problem of finding a consistent group of ASs which all together are able to solve a specific task. This set of ASs which are all together able to solve a specific task is called a *complete team*. Only tasks for which a complete team was found may be executed.

The ASs are separated into three groups:

$\mathcal{R}$ mobile manipulation systems (mobile robots),

$\mathcal{T}$ transport systems, and

$\mathcal{M}$ machining centers.

Mobile manipulation systems are able to work flexibly on the given tasks. Each time a mobile manipulation system finishes the execution of a task it can start to solve all the tasks it is able to no matter of its current configuration. At the best it is not necessary to change the configuration of the manipulation system, when a specific task has to be executed. In another situation it may be necessary to change the gripper but the manipulation systems has all the devices (e.g. the gripper needed) for the manipulation task with it. In the badest case it might be necessary to get some devices necessary for the manipulation task from a depot. The manipulation system will try to optimize the sequence of task execution by local criteria. Therefore, it will favour such tasks which it is able to execute using its current configuration.

For transport systems each transportation task may be split up into the subtasks: load, transport, and unload. It is possible to give to a transport system the transportation task as a whole task or to give the three subtasks to the transport system separately. When there is only one task specified for the whole transportation task, all the ASs involved in this task, the one to load and unload the transport system, are bound when the execution of the transportation task is started. If loading and unloading of the transport system needs much time these systems might be bound unnecessarily long.

Machining centers and flexible cells are much more restricted in their ability to choose tasks to be executed than ASs in the set $\mathcal{R} \cup \mathcal{T}$ because ASs in the set $\mathcal{M}$ are fixed to a location. An AS a in set $\mathcal{M}$ depends on ASs of set $\mathcal{R} \cup \mathcal{T}$ if not all of the devices needed for a specific task are already present within a. For this reason shall be assumed that the ASs in $\mathcal{M}$ are able to choose the next task to be executed purely by local criteria of optimality. In doing so, the configuration of the machining center or the flexible cell will be an important criterion to choose the next task to be executed. Nevertheless, the time constraints specified by the SFC system have to be noticed in the decision procedure.

Between the sets $\mathcal{R}, \mathcal{T}$ and $\mathcal{M}$ hold the precedence relations

$$\mathcal{M} \prec \mathcal{T} \prec \mathcal{R}$$

$\mathcal{T} \prec \mathcal{R}$ means that a member of $\mathcal{R}$ may only join a team for a specific task if all of the members of set $\mathcal{T}$ did already join the team for this specific task. The meaning of $\mathcal{M} \prec \mathcal{T}$ is analogous to $\mathcal{T} \prec \mathcal{R}$ with the difference that in each team only one AS belonging to $\mathcal{M}$ may participate. The precedence relation $\prec$ is transitive that means $\mathcal{M} \prec \mathcal{R}$ is valid too. The idea behind this definition is that the ASs which are able to execute tasks flexibly may react to the decisions of ASs which lack these flexibility in task execution.

To find complete teams, the ASs examine the list of tasks, which are announced by the SFC system, and try to reserve the task they would like to execute next for themselves. If an AS was able to reserve a task successfully for itself, this AS becomes the leader of the team for this task. The leader l of a team for a task t has to find a complete team for this task. l does this by sending messages to the other ASs which ask these ASs to join the team for task t. Because of the definition of the precedence relation $\prec$, it is clear that if an AS of set $\mathcal{M}$ can play a part in a specific task then it will always become the leader of the team for this task. If some ASs of set $\mathcal{T}$ can play a part in a task and no AS of set $\mathcal{M}$ can play a part in this task, then one of the ASs of set $\mathcal{T}$ will become the leader of the team. Only if all the parts of a task have to be played by ASs of set $\mathcal{R}$, one of these ASs will become the leader of the team for this task.

A conflict occurs if two team leaders send each other messages in which each of them asks the other one to join its own team. [5] describes a conflict resolution protocol for this situation which guarantees lifeness and fairness of the system.

2.4. The Task Planning Layer

The TCL of an AS passes a single task to the task planning layer (TPL) of the AS which has to be solved in order to solve the global task given on the TCL. The planning of the task is done using the content of a global knowledge base (see figure 5). This global knowledge base contains an image of the current state of the real world. By solving a task the ASs have to cooperate with each other. For that reason, the complex task given

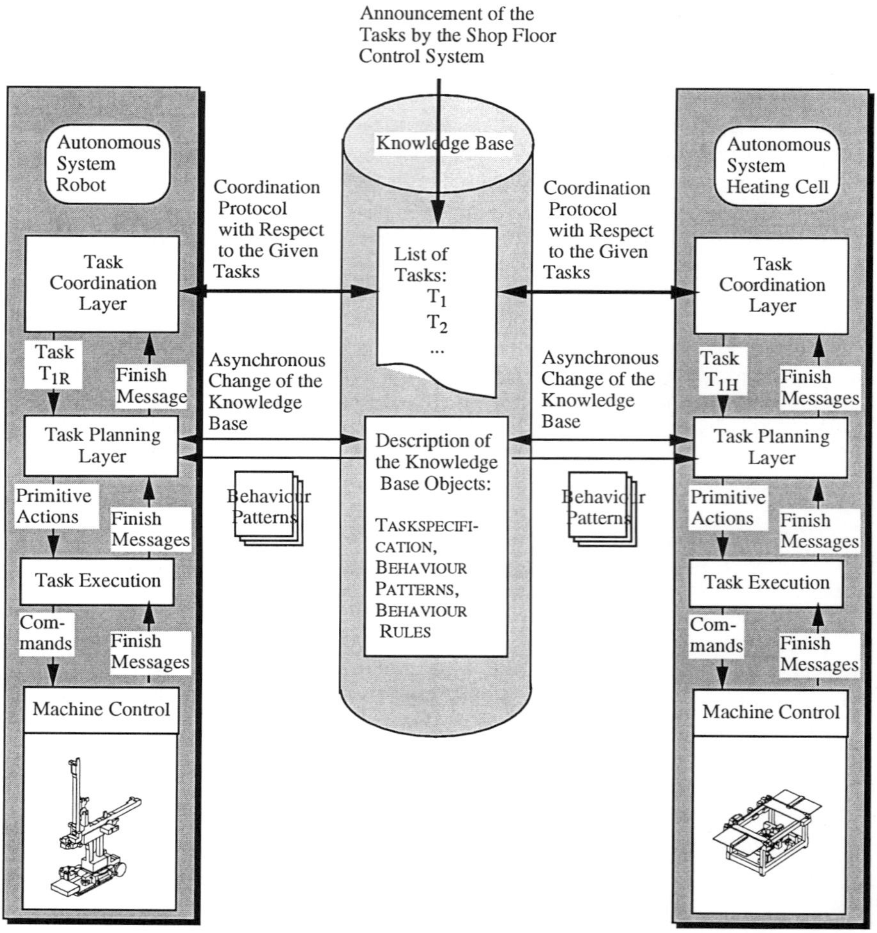

Figure 5. Description of the cooperation between ASs.

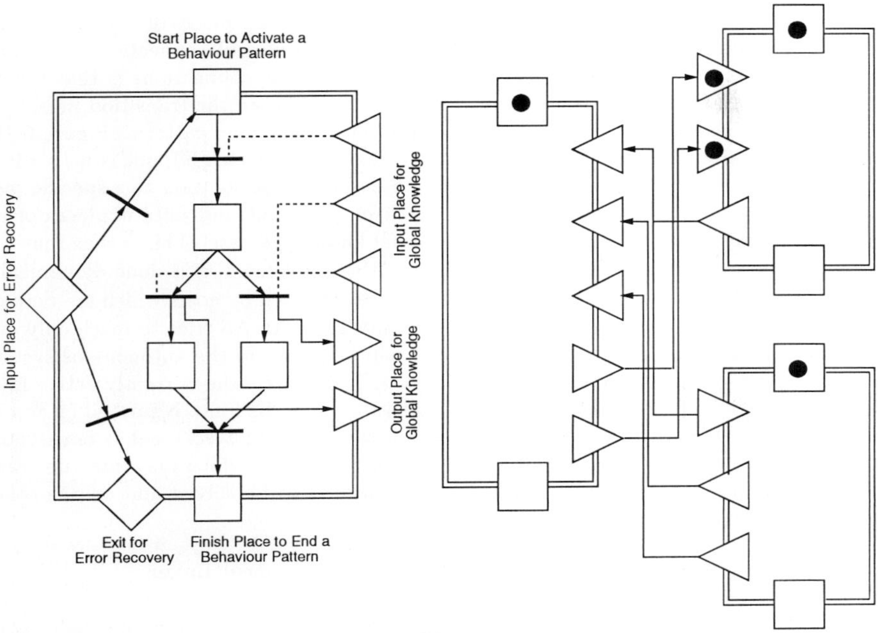

(a) (b)

Figure 6. EST net representation of behaviour patterns.

by the TCL has to be decomposed into primitive actions each of which can be executed by the AS without interacting with other ASs. The task planner contained in the TPL of an AS therefore implements the function:

$$f : \mathcal{K}^* \times \mathcal{T} \longrightarrow \mathcal{P}^*$$

where $\mathcal{K}$ is the set of a all possible knowledge base states, $\mathcal{T}$ is the set of all tasks, and $\mathcal{P}$ is the set of primitive actions. The input to the task planner is a sequence of knowledge base states because the content of the knowledge base is changed while task planning is active. Hence, the task planner reacts to changes in the environment of the AS by reacting to the changes in the knowledge base. For each task function f is specified by a set of rules which we call *behaviour pattern*. A single rule of a behaviour pattern is called a *behaviour rule*. [7] explains the strong connection between rule-based programming and predicate transition networks or, on a more abstract level, Petri Nets (place transition networks) [18]. For the description of behaviour patterns extended Petri Nets (EST nets) are introduced. These are coloured Petri Nets which may contain places of different types[1], which may contain inhibitor places, and which allow two different connections ($\longrightarrow$ and $—$) between places and transitions. The semantics of these connections is that tokens on this places are consumed or not consumed respectively when the transition fires.

Figure 6 (a) shows the EST net representation of a behaviour pattern. Figure 6 (b) shows that several EST nets may be linked together where each EST net is active in a separate AS. Therefore, the EST net for each AS describes the solution to a specific task from the point of view of this specific AS. Each of the input and output places of an EST correspond to an attribute value in the global knowledge base. This means that the connection between several EST nets of different ASs is not fixed, it is done dynamically unsing the global knowledge base. The EST nets are activated by goals which are derived from the tasks which are given from the TCL to the TPL. An AS tries to reach only one goal at a time and therefore only one EST net will be active in the autonomous system at a time. New goals may be derived by EST nets. In this case the currently active EST net is deactivated and a new goal which will activate a new EST net is pushed on top of the goal stack. Only when this new EST net is finished, the first EST net is reactivated again. Hence, the EST nets make up a hierarchy of behaviours which may interrupt each other, on one hand because errors occur while a task is to be solved and on the other hand because subtasks have to be solved within a task.

2.5. The Task Execution Layer and the Machine Control Layer

The primitive actions derived in the TPL of an AS are executed in the task execution layer (TEL) of the AS. By executing primitive actions, an AS has not to interact with other ASs. Nevertheless, such a primitive action itself may be a complex task, for example the sensor-guided insertion of a pin into a hole by a robot. The solution to such a task may be specified with the help of sensor/actor networks [3, 1].

Simple control commands are sent from the TEL to the machine control layer (MCL) where the actions are actually executed. The separation of the TEL and the MCL is done due to the systems which are available today. The controllers of these robots and machine tools provide only computational power and are therefore not able to execute complex programs. In the future it is likely that the controlling units of such systems will offer the computational power of today's workstations and that therefore the TEL and the MCL will grow together.

3. The Rule-Based Multi-Agent System MAGSY

The planning and controlling hierarchy described in section 2 was implemented based on the rule-based multi-agent system MAGSY [8, 6]. The kernel of an agent in MAGSY is a forward-chaining rule interpreter. Each agent in MAGSY is a self-contained entity with a unique identification. If an agent knows the identification of another agent, it is able to send this agent messages. In doing so the agents perform services and call on services of other agents. Agents are able to create new agents dynamically. If an agent creates a

[1] $\bigcirc$ places take up tokens (facts) which represent general internal knowledge of the agent, $\Box$ places take up tokens which represent goal facts, which trigger behaviour patterns, and $\lhd$ and $\rhd$ places take up tokens which represent global knowledge coming into or going out from a specific agent.

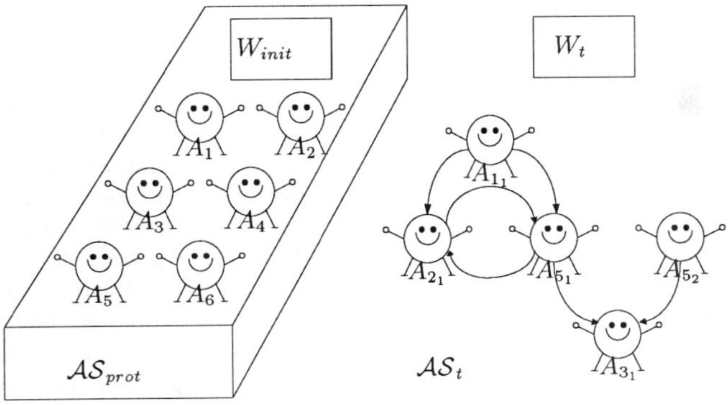

Figure 7. Static specification $\mathcal{AS}_{prot}$ of the solution for a problem and dynamic execution $\mathcal{AS}_t$.

new agent, it automatically knows the identification of this new agent.

The knowledge of the agents is structured in an object-oriented knowledge representation scheme. Objects are the basic elements which build a knowledge base. Objects which have the same structure are grouped together and make up classes. The description of a class is a prototype object which specifies how the elements (we call these elements instances or facts) look like. The properties of these instances are described by attributes. The classes are structured in a hierarchy in which attributes are inherited from the upper, more general, classes to the lower, more specific ones. There is a global knowledge base which contains the knowledge that may be accessed by all of the agents. Agents may store their identification in this global knowledge base and thus become known to all agents in the system.

For each of the planning and controlling components (PPC, SFC, TCL, TPL, and TEL) a MAGSY agent is introduced where the TCL, TPL, and the TEL agents are present for each AS separately[2]. Therefore, the term 'agent' which represents just an abstract computational unit having a specific structure must not be confused with the term 'autonomous system' which was introduced in section 2.3.

More formally, an agent in MAGSY is defined by a triple $(\mathcal{F}, \mathcal{R}, \mathcal{T})$, where

$\mathcal{F}$ is a set of facts which represent the local knowledge of the agent.
$\mathcal{R}$ is a set of rules which define the strategies for the general behaviour of the agent.
$\mathcal{T}$ is a set of services which are provided by the agent.

For a given agent A, $A_{\mathcal{F}}, A_{\mathcal{R}}$ and $A_{\mathcal{T}}$ denote the respective sets. Agent A may receive

[2]The MCL component of an AS is normally programmed in a standard programming language like PASCAL or C. It is therefore easily possible to integrate agents which are written in a imperative programming language into the MAGSY system.

messages which change the set of facts or activate a service of the agent. Each service of an agent consists of a set of rules which become active when the specific service has been activated by an incoming message. The execution of a service by an agent A may change the set of facts $A_{\mathcal{F}}$ and the set of rules $A_{\mathcal{R}}$. Therefore, the agents are learning when they execute their services, which can also mean that rules and facts can be forgotten, i.e. are removed. It is also possible that the execution of a service by agent A creates a number of new agents. The requests for services which are received by an agent are treated according to a specific strategy. Only one service can be active at a time. The other requests are queued and activated when the current service is finished. A priority is assigned to each service. A service s may be interrupted by a service i if a message arrives which activates i and i has a higher priority than s. If i is finished, s is reactivated. This process may happen recursively. Thus at any point in time, the service with highest priority is active. If more than one request for this priority class is present, the request belonging to the oldest message (that is the earliest one received) will be executed.

To model the solution of a given problem, a set of cooperating agents is specified. The static description of the solution of the problem is given by the prototype agent system $\mathcal{AS}_{prot} := (\mathcal{A}_{prot}, KB_{init})$, where

$\mathcal{A}_{prot}$ is the set of prototypical agents, the agents which may arise dynamically in the system. These agents are the potentially available problem solvers. Several instances of a specific prototypical agent can be created.

KB_{init} is the initial state of the global knowledge base.

The process of problem solving starts with the initial agent system

$$\mathcal{AS}_{init} = ((A_{1_1}, ..., A_{n_1}), KB_{init}),$$

where $A_{1_1}, ..., A_{n_1} \in \mathcal{A}_{prot}$. The computation goes on while the agents perform requested services and call on services of other agents. In doing so the agents cooperate by sending messages to each other. New agents may be created and killed again later on. A message may contain the identification of an agent, thus any desired structure of communication can be created dynamically. The contents of the global knowledge base can be accessed by all agents. Thus, the global knowledge base may be changed by the agents in any desired manner. For example, an agent can make its own identification known to all of the other agents by writing it into the global knowledge base.

In the example of figure 7 the system started with $\mathcal{AS}_{init} = ((A_{1_1}, A_{5_1}), KB_{init})$. A_{1_1} creates A_{2_1} and A_{5_2} and A_{5_1} creates A_{3_1}. At time t, A_{1_1} told A_{5_2} the identification of A_{2_1} and A_{5_2} told A_{2_1} its own identification. The identification of A_{3_1} came known to A_{5_2} because A_{3_1} stored its identification in the global knowledge base and A_{5_2} extracted this information from the global knowledge base. In this example there is no other way for A_{5_2} to get the identification of A_{3_1}.

4. Conclusion

By this paper a hierarchical planning structure for a flexible manufacturing system was presented. The hierarchical planning structure consists of the layer of the production planning and control (PPC) system, the layer of the shop floor control (SFC) system, the task coordination layer (TCL), the task planning layer (TPL), the task execution layer (TEL), and the machine control layer (MCL).

For the layer of the SFC a concept for knowledge-based, reactive scheduling was described. The scheduling is done at hand of a statistical model which is extracted from the actually present data. Because of this there is only a loose coupling of the SFC and the autonomous systems (AS) which are controlled by this SFC, the ASs do local optimization with respect to their current situation.

In the hierarchical planning structure an autonomous system (AS) is modelled as self-contained unit having a hierarchical planning structure itself consisting of the four layers: TCL, TPL, TEL, and MCL. Task planning in the TPL of an autonomous system is done using behaviour patterns which were described briefly in the paper.

The hierarchical planning structure has been implemented for the model of a manufacturing plant [4, 7]. The implementation was done with the help of the rule-based multi-agent system MAGSY [8, 6], which allows distributed parallel rule-based programming. The abstract computation model of MAGSY is described at the end of the paper. Experience with the implementation of hierarchical planning and controlling structure showed that the planning and controlling components of the physical devices of the ASs where events arise asynchronously can, on a symbolic level, be easily described by these concepts. The forward-chaining reasoning semantics of the rule interpreter which is built into a MAGSY agent can be used here optimally. With the ideas presented by this paper it is possible to describe reactive behaviour for ASs on a high symbolic level. Future research will now try to enhance this planning paradigm to make it possible for the ASs to do more powerful deliberative planning than it is possible till now.

REFERENCES

1. Rodney A. Brooks. A Robust Layered Control System for a Mobile Robot. In Patric Henry Winston and Sarah Alexandra Shellard, editors, *Artificial Intelligence at MIT, Expanding Frontiers*, pages 3–27. MIT Press, Cambridge, Massachusets, 1990.
2. Anke Carstensen. Ein Konzept zur wissensbasierten Abwicklung des internen Materialflusses in einer flexiblen Fertigungszelle (A Concept for the Knowledge-Based Control of the Flow of Material in Flexible Cell). Master's thesis, TU München, Institut für Informatik, November 1990.
3. Jonathan H. Connell. *Minimalist Mobile Robotics*, volume 5 of *Perspectives in Artificial Intelligence*. Academic Press, Inc., 1990.
4. Klaus Fischer. Concepts for an Agent System in a Flexible Manufacturing System. In *Proc. of the 23rd Int. Symp. on Automative Technology and Automation*, Vienna, December 1990.
5. Klaus Fischer. Rollenverteilung unter gleichberechtigten Agenten (Role Casting in Non-hierarchical Agent Societies). In *Beiträge zum Gründungsworkshop der Fachgruppe Verteilte Künstliche Intelligenz*, DFKI, Saarbrüecken, April 1993.
6. Klaus Fischer. The Rule-based Multi-agent System MAGSY. In *1992 Proceedings of the Special Interest Group on Cooperating Knowledge Based Systems*, DAKE Centre: University of Keele, 1993.
7. Klaus Fischer. *Verteiltes und kooperatives Planen in einer flexiblen Fertigungsumge-*

bung (Distributed and Cooperative Planning in the Environment of a Flexible Manufacturing System). DISKI, Dissertationen zur Künstlichen Intelligenz. infix, 1993.
8. Klaus Fischer and Hans-Michael Windisch. MAGSY: Ein regelbasiertes Multiagentensystem (MAGSY: A Rule-based Multi-agent System). *KI*, 1/92, 1992.
9. M. Fox. *Constraint-Directed Search: A Case Study of Job-Shop Scheduling*. PhD thesis, Carnegie Mellon University, Pittsburgh, USA, 1983.
10. Axel Groha. *Universelles Zellenrechnerkonzept für flexible Fertigungssysteme (Universal Concept for a Cell Controler in a Flexible Manufacturing System)*. Springer-Verlag, 1988.
11. Wolfgang Haunschild. Untersuchung über die Anwendbarkeit von Optimierungsstrategien auf die Auftragsreihenfolge eines Zellenrechners unter Echtzeitbedingungen (Analysis of the Applicability of Optimazation Strategies on the Job Sequencing of a Flexible Cell with Respect to Real-time Constraints). Master's thesis, TU München, Institut für Informatik, Mai 1990.
12. G. Hirschberg. Produktionsautomatisierung und Flexibilität aus der Sicht des Informatikers (Automation and Flexibility from the Point of View of Computer Science). *VDI Bericht NR. 723*, 1989.
13. E. Horowitz and S. Sahni. *Algorithmen — Entwurf und Analyse (Algorithems — Design and Analysis)*. Springer, Berlin, 1981.
14. Elisabeth Jelinek. Implementierung einer flexiblen Fertigungssteuerung am Beispiel eines fischertechnik-Modells mit Hilfe von KI-Werkzeugen (Implementation of the Control of a Flexible Manufacturing System at Hand of the Example of a fischertechnik Model using AI Tools). Master's thesis, TU München, Institut für Informatik, November 1991.
15. Sabine Koenig. Entwicklung von Dispositionsstrategien für ein Fertigungsleitsystem (Development of Scheduling Strategies for a Shop Floor Control System). Master's thesis, TU München, Institut für Informatik, November 1990.
16. Thomas Kupec. Integration of Autonomous Mobile Robots in Flexible Manufacturing Systems. In *Proc. of the 2nd Inter. Conf. on Intelligent Autonomous Systems*, pages 122–133, Amsterdam, December 1989.
17. H. Müller-Merbach. *Optimale Reihenfolgen (Optimal Sequences)*. Springer Verlag, 1970.
18. Wolfgang Reisig. *Petri Nets*, volume 4 of *EATCS Monographs on Theoretical Computer Science*. Springer-Verlag, 1985.
19. Norman Sadeh. *Look-ahead Techniques for Micro-opportunistic Job Shop Scheduling*. PhD thesis, School of Computer Science, Carnegie Mellon University, CMU-CS-91-102, March 1991.
20. A.-W. Scheer. *CIM — Der computergesteuerte Industriebetrieb (CIM — The Computer Controlled Industrial Concern)*. Springer Verlag, 1990. 4. neu bearbeitete und erweiterte Auflage.
21. Katia Sycara, Stephen Roth, Norman Sadeh, and Mark S. Fox. Coordinating Resource Allocation in Distributed Factory Scheduling: A Constraint Heuristic Search Approach. *IEEEE EXPERT*, 6(1):29–40, 1991.
22. Hans-Jürgen Zimmermann. *Operations Research Methoden und Modelle (Operations Research Methods and Models)*. Vieweg, 1992.

Knowledge-Based Reactive Scheduling (B-15)
E. Szelke and R.M. Kerr (Editors)
Elsevier Science B.V. (North-Holland)

APPLICATION OF NEURAL NETWORK PROCESS MODELS IN REACTIVE SCHEDULING

B. J. Garner and G. J. Ridley

Department of Computing and Mathematics, Deakin University, Geelong, Victoria 3217, Australia

Abstract

Preliminary study of neural network process models in scheduling production for glass coating operation (AIRCO process) has identified knowledge rich control strategies for reactive scheduling. Substantial progress in the data engineering phase of the project has resulted in new insights into the knowledge acquisition requirements relevant to neural network design.

Keyword codes: I.6.5; I.2.11
Keywords: Simulation and Modelling, Model Development; Distributed Artificial Intelligence.

1. INTRODUCTION

In this paper we are concerned with maintaining control over an environment where the predetermined schedule of jobs can be interrupted. Reasons for interruptions include: urgent orders of replacement glass, a suspension to coating to conduct minor maintenance such as a burn-in, or a reassignment of job priorities.

The control problem arises from the different levels of complexity associated with different coated products and with the selection of an optimal job mix that reflects set-up requirements and coating reproducability arising from specific coating sequences.

The use of neural networks in the control of coating reproducability requires the selection of training sets appropriate to each combination of circumstances. The reactive scheduling issue, therefore, resolves itself into two selection problems : a) identification of the best neural network parameters; and b) identification of the appropriate training set.

2. OUTLINE OF THE GLASS TINTING PROCESS

Pilkington's Dandenong plant uses an Ion-implantation method to tint architectural glass. The Ion-implantation is achieved using a vacuum technology that is proprietary to the AIRCO company. The coating system is a continuous process in which the glass enters a series of high vacuum coating chambers.

AIRCO uses a sputtering process to deposit thin films of metal atoms on the glass. Sputtering is the bombardment of a target, the material to be deposited, by ions which physically dislodge atoms off the target (the cathode), causing them to be ejected from the surface and to strike the glass surface and adhere to it.

The coating zones contain the specially designed cathodes and the process gas. The sputtering sources, which are planar magnetron cathodes, contain the target material such as titanium, silver, stainless steel or inconel that will be implanted into the glass surface. A plasma ignites within the vacuum chamber when the appropriate conditions are achieved. The gas is ionised by the electric field and positively charged atoms of this gas (ions) are attracted to the surface of the target, which is negatively charged. The process gas can be either inert or reactive, which allows not only pure elements, but also compounds to be deposited at the glass surface.

2.1. Characteristics of coated products

Pilkington's produce a range of products coated to give different optical and solar performance characteristics, along with a variety of colours. Apart from reflective glass, the AIRCO process can also make mirrors, using metals other than silver, and low-emissivity glass which has been especially developed to retain heat in buildings in cold climates.

The control of these coatings ranges from relatively easy ones which require minimal online adjustments by the operators, through to coatings which have been discontinued due to the difficulty in ensuring reproduceability.

The initial state of the chambers is never the same on any two occasions. This is due to cathode wear and the changing base pressure of the coating zones. Operators must determine a new set of control settings each time a new batch of glass is to be coated.

One of the major influences on the colour of the coated class is the thickness of the coatings. The thickness of the coating is effected by the speed at which the glass travels through the coating chamber as well as the sputtering rate of the coating process. The sputtering rate is the speed at which a material is actually deposited onto a substrate. The sputtering rate is dependent on the number of atoms of target material removed by the bombardment of the ionised gas particles (known as sputter yield). Factors that effect the sputter yield and the resulting deposition rate include the mass of the target atoms, the mass of the gas ions, the energy level of the gas ions, and the rate of emission of the secondary electrons from the target as it is bombarded by the gas ions (AIRCO, 1989)

In short, sputtering occurs at different rates because different cathode materials have different masses, different bonding strengths, and emit secondary electrons at different rates. For example silver, gold and copper have relatively high sputtering yields when compared to Titanium, and thus higher power levels are required for Titanium to achieve the same yield as Silver.

The gases used in the sputtering process are either inert or reactive gases. Inert gases include helium, neon and argon, while reactive gases include oxygen, hydrogen, and nitrogen. If an inert gas, such as argon, is used to sputter a titanium target, then a pure metallic titanium coating will be deposited. Alternatively, if a reactive gas, such as nitrogen is bled into the coating chamber then a titanium nitride coating is produced.

The type of gas, the rate of flow, the partial pressure will determine whether the sputtered coating is a metal, an oxide, or a nitride. These and the other process variables such as linespeed, cathode material, and cathode wear, effect the coating characteristics such as colour, transmission, reflection, light absorption and durability of the coating.

The process is affected by many parameters, some variable and controllable during a process run, others are fixed. Dynamic control of the process is achieved by varying the power to each cathode, the line speed, the type of gas in the coating zones and the gas flow rates. Target wear and changing base pressures are the cause for adjustments in initial control settings.

The quality of the coating is very sensitive to the initial set-points determined by the operators. The early detection and correction of process drift is crucial for the success of a production run.

3. THE CONTROL PROBLEM

Many different types of coatings are produced by Pilkingtons. There are families of coatings which are made using the same materials, and the different optical characteristics are achieved by varying the thickness of the coatings. If, for example, the coatings known as TS20, TS30, and TS40 (which are all single layer titanium nitride coatings with additional nitrogen in their composition), are to be processed during the same day, then it would be appropriate to schedule these jobs together as they require similar conditions in the coating chamber. The additional information we hope to provide is what order should these three be scheduled, based on each coating's sensitivity to cathode wear. If for example the TS40 is more susceptible to cathode wear than the TS20, then obviously, the TS40 coating should be scheduled before the TS20.

When both titanium and oxygen are used in a coating chamber, an unwanted build up of $Ti0_2$ on the surface of the cathode will result. The $Ti0_2$ can be burnt off by applying a large amount of power to the cathode. Coatings that use both titanium and oxygen should therefore be scheduled last, as burn-offs can then be conducted after the production run.

Orders can be divided into two basic groups. Those which have been generated by specific customers and those which are stock plates. Stock plates are used within the company, or sold to other companies, to be used in particular products such as double glazed units. Stock plates are relatively basic coatings, which are reasonably stable and easy to control (because operators have plenty of experience with them). The special orders can be more elaborate coatings (requiring multiple layers of material), and thus harder to control. These coatings may be harder to control because the operators have had less exposure to these coatings.

A coating can also be interrupted if a replacement sheet is needed for a previous coating. This can result from a breakage during the post-processing of the glass. These interruptions may be unavoidable because the replacement sheet is preventing the delivery of a large order.

The operators do not receive any immediate results from their control actions. This is because the glass is measured after it leaves the coating process. Because of the time delay in feedback from any process adjustments, the operators often try to save time by adjusting more than one parameter at a time. This can result in confusion as to which control actions are responsible for particular effects. The operators can use the interstage transmission monitors to get an idea how the coating is proceeding. The interstage transmission monitors are located after each of the three coating zones. They measure the percentage of white light passing through the glass, and do not measure the colour or the glass.

It is important to note that the interstage transmission monitors are a limited form of feedback during the coating process. The monitors slowly deteriorate as a result of their protective viewing window being tinted by the coating.

The Spectrophotometer measures the colour of the glass at the end of the tinting process. The resulting colour co-ordinates of the glass, which are the Hunter Lab values, are displayed to the operators The operators will adjust the set-points of the process if the colour is drifting outside the tolerance bands. The type of control action, and the tolerance bands vary depending on the coating being done.

4. DATA ENGINEERING RESULTS

The data acquisition system collects both input and output data from the coating process. The input data is collected from the linespeed controller, each of the 9 cathode power controllers, and from the 6 gas flow rate controllers. The entry and exit valves are also monitored so that the number of sheets inside the coating chambers can be determined.

The output data is collected from the spectrophotometers host computer. The host computer contains the following information :-

- Time and date of measurement
- The name of the standard (the desired colour of the coating).
- A unique identifier which is either the batch or the customer order number.
- The desired Hunter L, a, and b colour co-ordinates that characterise the coating
- The desired spectral curve at 20 nm intervals from 400nm to 700nm inclusive.
- The measured spectral reflectance data measured at 20nm intervals from 400nm to 700nm inclusive.
- The L, a and b values calculated from the spectral reflectance data.

Fig 1 shows some typical results that were obtained using the data acquisition system . The coating being conducted was a TS30 on clear glass which has a desired Hunter L, a, and b values of 36.94, -1.28, and -10.06 respectively. The time scale is in minutes since midnight.

The graph of the power levels on each cathode illustrates that initially a burn-in was conducted followed by a series of line-ups. The adjustments by the operators during the line-ups have improved the Hunter L, and the Hunter a values. The Hunter b values have degraded slightly. The production run has then commenced at time 910 minutes. An adjustment to the cathode power levels has been carried out approximately 30 minutes after

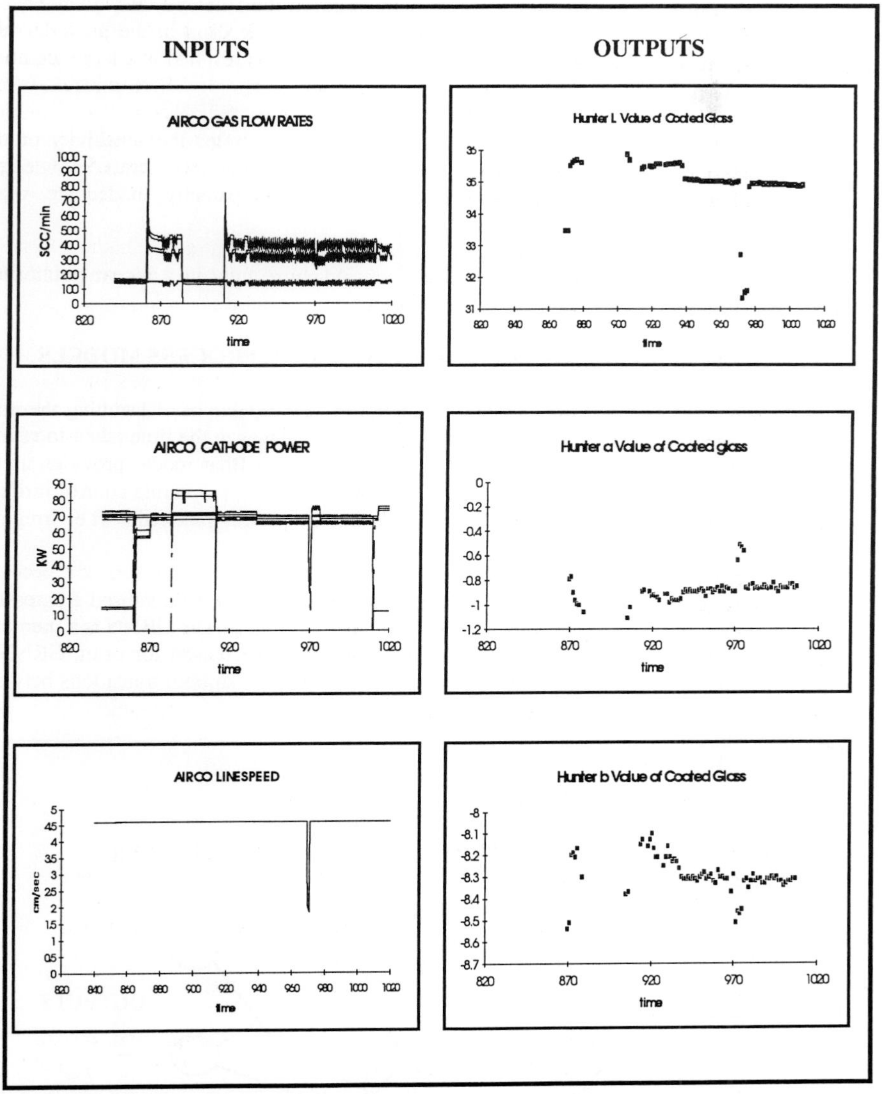

FIG. 1 Graphs of input and output data collected during TS30 coating on 16/3/1993

the commencement of the production run, which has reduced the Hunter L and b values, but had little effect on the Hunter a values of the coated glass. The drop in the power to the cathodes and the drop in linespeed at around the 970th minute has resulted in a large variation in the colour.

The variability of the outputs (ref. Figure 1) clearly demonstrates the sensitivity of the coatings to operator settings and manual intervention. The operators "craft knowledge" ensures satisfactory outcomes for standard products, but is patently inadequate when developing new products or handling unusual setups.

The scope of the data engineering stage is discussed more fully in a recent publication (Garner et al, 1993)

5. COMPARATIVE ANALYSIS OF NEURAL NETWORK PROCESS MODELS

Back propagation has been shown (Bhat et al,1990) to be capable of of handling the non-linearities in spectral data, but there are considerable disadvantages in the time taken to retrain such networks and there is no guarantee of convergence. The Bhat model provides useful insights, however, in the use of artificial neural networks (ANN) for dynamic control through the study of disambiguation of desirable and undesirable spectral properties (i.e. irregular features) .

The scheduling advisor selected for investigation is based on the Generalised Regression Neural Network (GRNN) developed by Donald Specht (refer Fig. 2).The GRNN is a memory-based neural network, that has a one-pass learning algorithm.The reason for using GRNN is the speed at which a continuous model can be constructed and the smooth transitions between observed values.

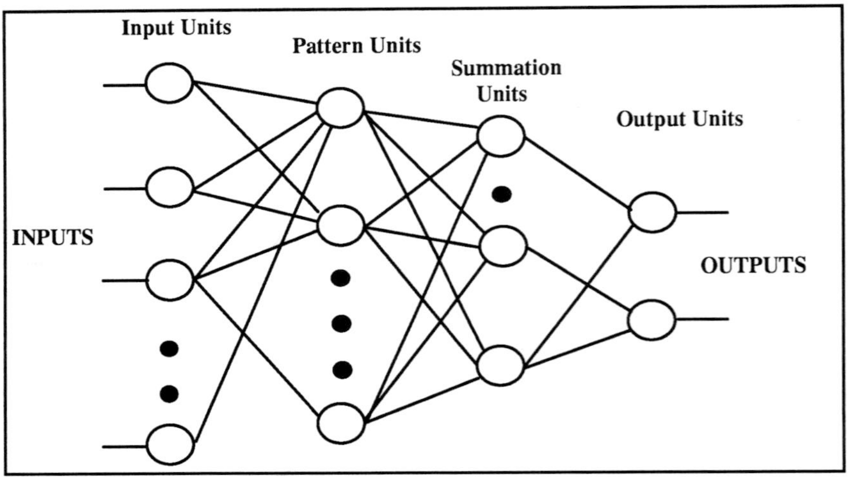

Fig. 2 The Generalised Regression Neural Network

GRNN models may also be desirable control aids in engineering a total, reactive,control system.It is, therefore, appropriate to review the knowledge requirements for control.

6. KNOWLEDGE-BASED ADVISOR FOR REACTIVE SCHEDULING

An outline of the knowledge-based supervisor is shown in Figure 3. The rule base (KB) provides supervisory control (reasoning engine) when discontinuities are encountered in the response surface, and it is necessary to select an alternative ANN model for control of the process. The rule base, in conjunction with the cumulative database, also provides advice to the operators on the set-up changes required when production plans are interrupted or unexplained conditions emerge.

The KB and the massive database collected over time during AIRCO coater operations provide a powerful, dynamic control capability in support of the ANN facility. For example, they support determination of;

- Control settings used in previous coatings;
- Settings that resulted in the best coatings;
- Cathode useage between coatings or since a particular coating;and
- Current coating performance in comparison to previous coatings. This will help to determine any corrections that may be necessary.

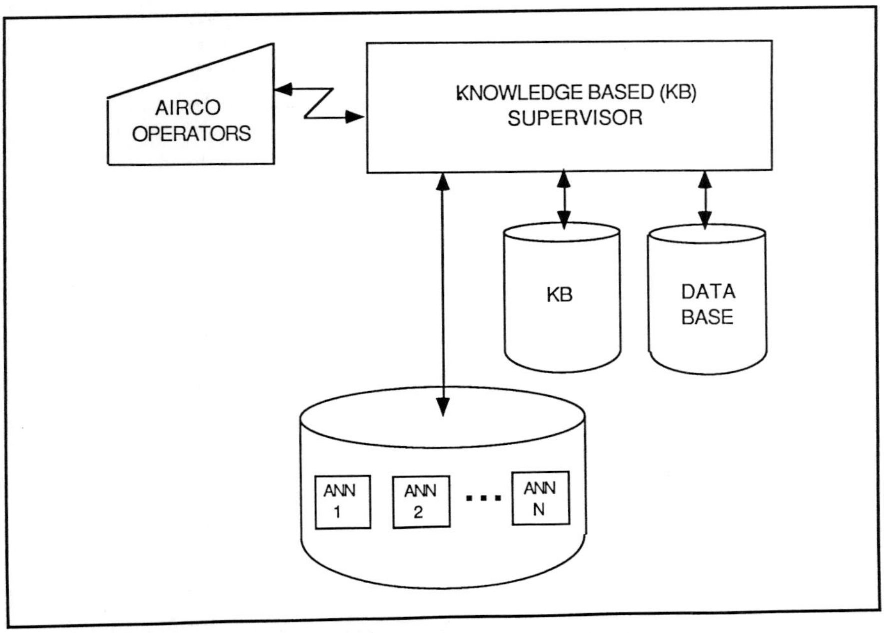

Figure 3 Expert Advisor for Production Scheduling

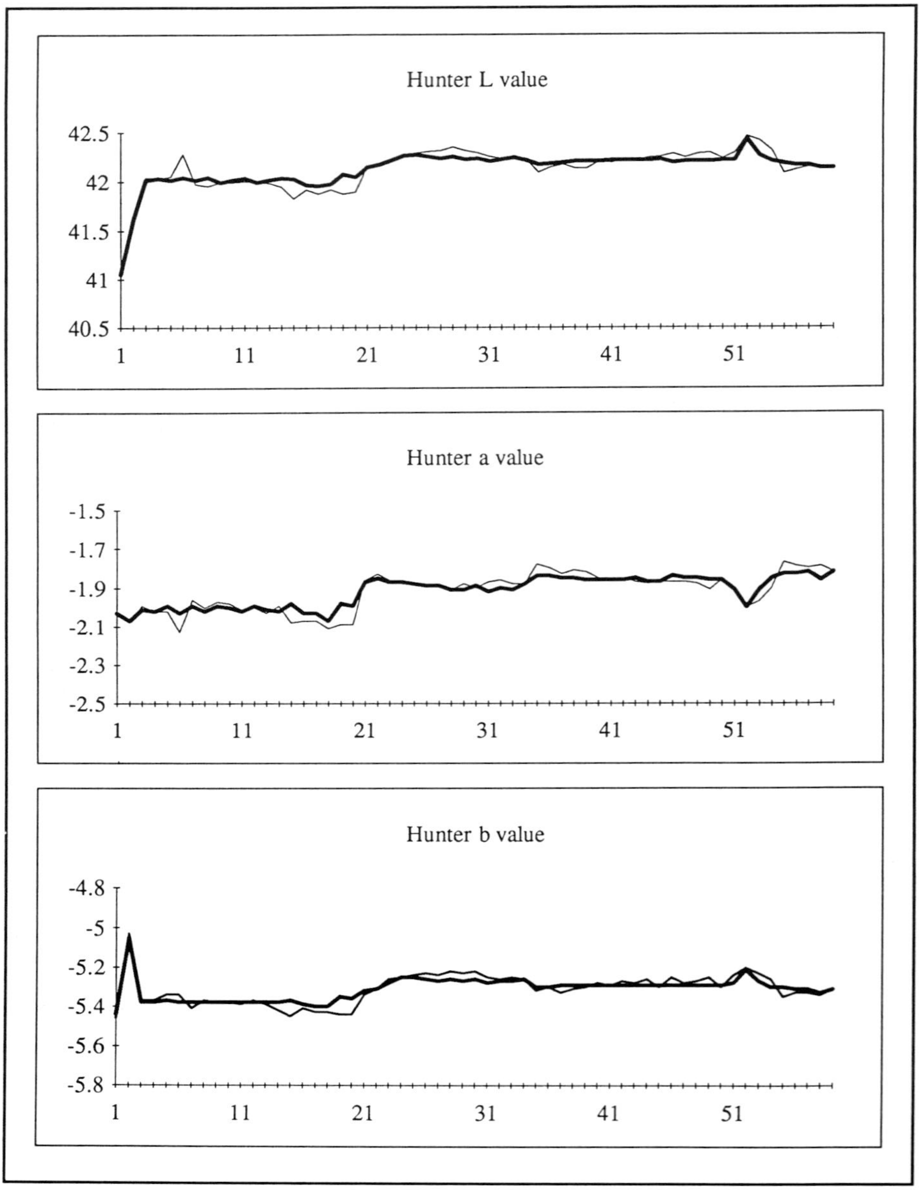

Figure 4. Output of GRNN (thick line) and of the AIRCO process (thin line).

The use of GRNN's as the artificial neural networks enables a new process requirement or operational model of the current coating process to reflect changes required to the current operating parameters, or by reference to the database, to establish initial settings for a new batch.

7. PRELIMINARY RESULTS USING GRNN MODELS

Data from two production runs of a TS21 coating have been used to train a GRNN. The GRNN was trained using just 20 exemplars from a previous coating and the first 10 exemplars from the second coating. The GRNN was then used to predict the following 30 Hunter Lab values. The results appear in Figure 4.

8. CONCLUSIONS

A production scheduling adviser has been outlined that uses a knowledge base, and a database of past and present coating data, to construct a neural network that models the current state of the AIRCO process. The one-pass learning algorithm of the GRNN enables a model of the system to be constructed quickly. This model can then be used to help the operator maintain control over the coating environment or establish initial settings of the AIRCO machine.

ACKNOWLEDGEMENTS

This research is supported by Pilkington Australia and the Australian Research Council APRA Scholarship (Industry) Grant Number PG90/342.

REFERENCES

AIRCO, (1989); Process Theory and Practice GR-20, AIRCO operations manual, AIRCO Coating Technology, U.S.A, 1989, Chapter 4

Barto, A.G.(1991), Connectionist Learning for Control: an Overview, Neural Networks for Control, Chapter 1 ,Miller,W.T.,Sutton, R.S., and Werbos, P.J., MIT Press 1991.

Bhat, N.V., Minderman, P.A., Jr, McAvoy, T,and Wang, N.S.,(1990), Modelling Chemical Process Systems via Neural Computation, IEEE Control Systems Magazine, April 1990, pp. 24-29.

Garner, B.J., Ridley, G.J., and Lowe, P.J, (1993), Data Engineering for Neural Net Analysis of Glass Furnace Characteristics. Proceedings of the ANNES conference, New Zealand, November 1993.

Narendra, K.S., and Parthasarathy, K,(1990) Identification and control of dynamical systems using neural networks, IEEE Trans. Neural Networks, March 1990, pp. 4-27.

Specht, D.F., (1991) A General Regression Neural Network, IEEE Transactions on Neural Networks, November 1991, pp. 568-576.

Knowledge-Based Reactive Scheduling (B-15)
E. Szelke and R.M. Kerr (Editors)
Elsevier Science B.V. (North-Holland)

Production Activity Control for Small Manufacturing Enterprises

Paddy Jordan, Jim Browne, Mary Browne

CIMRU,
Department of Industrial Engineering and Information Systems
University College Galway, Ireland

Abstract: Often the production management strategies and solutions applicable to large manufacturers may not be suitable for smaller companies. This paper describes an architecture for Production Activity Control and shows how this may be adapted to suit the needs of Small Manufacturing Enterprises (SMEs). In many SMEs the human scheduler is the main source of expertise and knowledge within the Production Activity Control system. We recognise the need for a framework within which the human expert can exert control.

KEYWORDS: Production Activity Control, Scheduling, Decision Support, Small Manufacturing Enterprise.

Small Manufacturing Enterprises (SMEs) are playing an increasingly important role in the Irish economy. At the beginning of 1992 more than 80% of the companies in Ireland employed less than 50 people while less than 3% of the 7400 companies had more than 500 employees [8]. We see a SME as being a company which employs up to 100 people. SMEs tend to exhibit some differences from larger companies. Cross-functional management is not unusual. SMEs tend to have single-cell production systems, and the products produced by SMEs would tend to be less complex in structure than those produced by their larger counterparts. Although this may, in many cases, mean that the manufacturing system is less complex, it must also be remembered that SMEs are subject to certain problems such as financial constraints and an absence of technological expertise.

Here we propose an approach to shop floor control for SMEs. There are three main parts to the paper. These are:

1. Defining an architecture for Production Activity Control (PAC) in SMEs.

2. Providing the tools, within the PAC architecture, to allow the human scheduler to react to unexpected events in the production system.

3. Providing decision support for the human scheduler.

The PAC architecture aims to improve control over the manufacturing system thus increasing the predictability of the system. However, since no manufacturing system is totally predictable, the architecture allows the freedom to utilise the knowledge of the human scheduler in reacting to unexpected events.

1. An Architecture for Production Activity Control

Duggan and Browne [3] outline an architecture for Production Activity Control which is aimed at reducing the gap between shop floor control theory and practice. The architecture breaks down the task of PAC into five major building blocks: a scheduler, a dispatcher, a monitor, a producer and a mover. The *scheduler* creates a short term plan for production. It accepts the production requirements from a higher planning system (production planning) and develops a detailed plan which determines the precise use of the different manufacturing facilities within a specified time frame. The PAC scheduler takes the schedule guidelines from the Factory Coordination[1] system and develops a plan that can be released to the shop floor by means of the PAC dispatcher. The scheduler represents the planning aspect of PAC, a schedule is developed taking different variables and constraints into account. However, once this schedule is released to the shop floor it becomes susceptible to unexpected events. It is the role of the dispatcher to deal with these unexpected events which threaten the validity of the schedule.

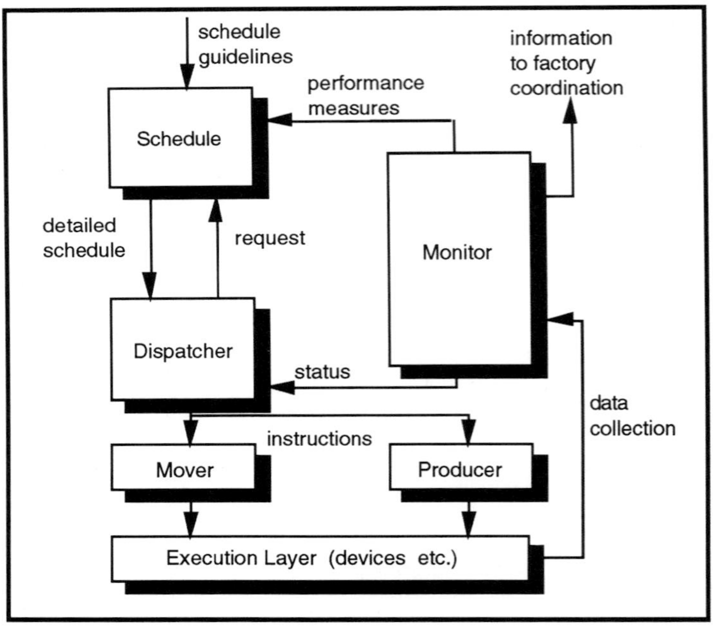

Figure 1: Production Activity Control Architecture [3].

[1] The Factory Coordination system organizes the flow of products throughout a factory and ensures production in each cell is synchronized with the overall production goals of the factory.

The *dispatcher's* role is to react to the current state of the shop floor and choose the best possible course of action. It is the real-time implementation of the schedule. The dispatcher relies heavily on accurate and timely data. When a decision has been made the next step is the production process, the dispatcher sends instructions to both the mover and producer to execute these steps.

The *mover* coordinates the material handling function and interfaces between the dispatcher and the physical transportation and storage mechanisms on the shop floor. The *producer* is the process control system within PAC which contains or has access to all the information required to execute the various operations at a workstation.

The *monitor* provides status information to both the scheduler and the dispatcher. For example, the monitor may report the WIP levels from the shop floor, or the availability of a machine. Higgins [4] identified the three main activities of the monitor as data capture, data analysis and decision support. The data capture system collects data from the shop floor. This is then translated into information by the data analysis system and can then be used as decision support for the various PAC activities. This traditional PAC architecture is shown in figure 1.

1.1 Adapting the PAC architecture for SMEs

According to Mendonça et al [7] there is no proven framework for low cost implementation of shop floor control systems for small and medium enterprises yet available. The existing SME shop floor control systems are partial solutions developed for specific environments and objectives. The PAC architecture described above is particularly suitable for large, highly automated, manufacturing systems. There are many differences between these large manufacturing organisations and SMEs. Some of these are listed below. These differences are seen as typical but not absolute and are based on practical experience.

- Large manufacturing organisations tend to be MRP-driven in that they will tend to have a Master Production Schedule (MPS) detailing the aggregate production requirements for a long term period and will use this to plan both short-term production and materials requirements. On the other hand, SMEs will tend to produce a less complex product by nature, often have a shallow Bill of Material and tend to be customer order driven.

- Large manufacturing organisations tend to fit into the categories of Make-to-Stock and Assemble-to-Order while SMEs tend to Assemble-to-Order and often Make-to-Order.

- Large manufacturing organisations tend to have a hierarchical organisational structure with a number of levels between top management and the operators on the shop floor. Smaller enterprises have a *flatter* organisational structure as they have fewer employees and as a result management cover a broader range of activities than their counterparts in a large firm.

- Large manufacturing organisations tend to have a high degree of product variation and are thus often multi-cellular while SMEs more often contain a single cell and produce a narrower variety of products. Product variation in small firms is a result of highly flexible equipment which can be adapted or modified to suit product demand. In larger firms with less flexible equipment, variation is catered for by setting up cells to produce diverse products.

- Large organisations may have the need and the resources to be highly automated whereas SMEs may not be able to justify the expense of automation with respect to their size and turnover.

To adapt the PAC architecture for SMEs we must take account of these differences. A representation of the modified PAC architecture is illustrated in figure 2.

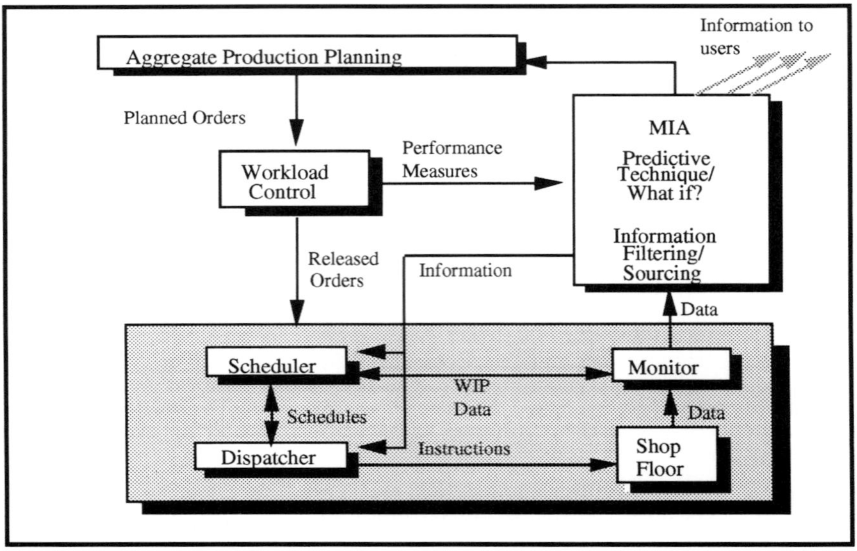

Figure 2: Production Activity Control for Small Manufacturing Enterprises.

Aggregate Production Planning

Because SMEs tend to be order driven we cannot assume that MRP output is available to the PAC system. We therefore introduce the idea of an *aggregate production plan* which is a forecast of likely production over some future period. The aggregate production plan represents the highest level of planning in a SME. Aggregate production planning is the control process that specifies the volume of production for a given period. It has to anticipate potential difficulties with respect to capacity availability and the availability of items with long lead times [2]. The results of the aggregate production planning process are firstly, a delivery plan, which states planned deliveries for a given period. Secondly, a capacity plan, which states the amount of production hours required for given resources. Thirdly, an inventory plan which specifies the required inventory for various production stages for a given period.

Workload Control

A system based purely on the PAC architecture could allow fluctuations in load to the shop floor. In an Assemble-to-Order/Make-to-Order system an attempt should be made to minimise these fluctuations and balance production across the scheduling period. To do this we introduce the concept of workload control which balances the amount of work-in-progress in a cell.

For all work orders which have been processed by critical capacities (potential bottlenecks), the amount of required capacity is calculated per critical capacity. The total amount of capacity that is required from a critical resource is termed the remaining workload of that resource. Norms are set for remaining workloads of critical capacities. Remaining workload norms reflect desired throughput times. Desired throughput times are set by making a trade off between resource utilizations and the throughput time that can be achieved with different utilization levels. Figure 3 illustrates how the remaining workload of the critical resources are maintained at a certain level by means of controlling the release of work orders.

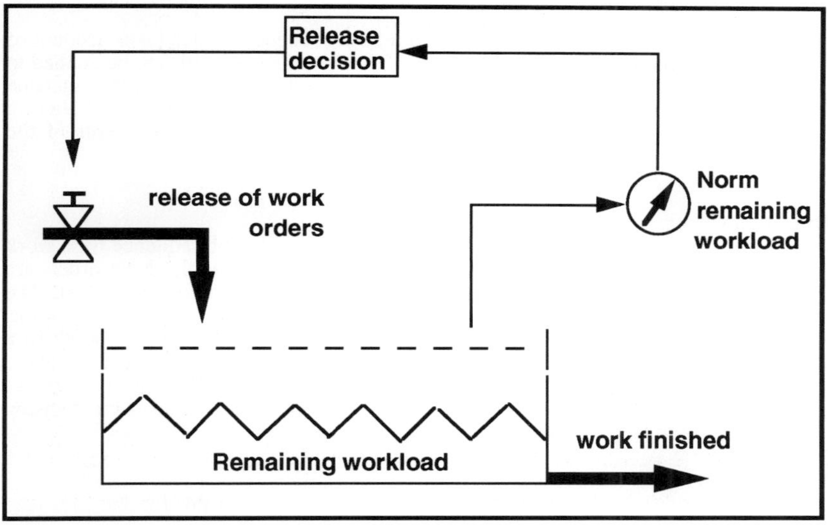

Figure 3: Workload Control [2].

There is an explicit work order release decision for a resource in a workload controlled situation. Every time a short term production plan is made, release restrictions are set for critical capacities. The objective is to come to a stable situation at the end of a release period. For each critical capacity, the available net capacity in this release period is calculated. Then, for each release restriction, remaining workload deviations from the remaining workload norms are extracted and added to the available net capacity. The output of workload control is a set of release restrictions for its critical resources. This is then used to release particular work orders to the scheduler.

Many SMEs are single cell facilities and as such use workload control as a mechanism for controlling the release of orders to the shop floor with the objective of keeping the remaining workload for bottleneck resources between certain predefined norms. As SMEs tend to be customer order driven, the value of applying workload control increases. This is due to the fact that the predictability of the lead time increases in a workload control situation and therefore workload control provides a better basis for delivery date adherence in a made-to-order type situation.

The Scheduler

The Scheduler in Production Activity Control for SMEs is used to plan start times for each order released by workload control on each operation in the routings for the respective products. The tasks of the scheduler are as follows:

- Accept the orders as input from the Workload Control element.
- Generate a realistic first pass schedule.
- Provide an ability to alter the schedule to account for unexpected events in the manufacturing system or to impose the schedulers knowledge on the system.
- Release the schedule to the dispatcher so that it may be implemented.

The scheduler should be able to generate a realistic schedule which takes account of the system capacity and constraints and allow for scheduling heuristics to be applied to each operation within the route. In this way the human scheduler is spared the mundane task of generating a first pass schedule and can thus concentrate on the task of altering the schedule so that it becomes tailored to the more specific requirements of the manufacturing system.

The Dispatcher

The *dispatching* task involves the final determination of the job sequence for a work centre. It is essentially the *real time* implementation of the schedule. After orders are released to the shop floor, they must be controlled from one operation to the next. The dispatching activity is responsible for sending information on the start of PAC works orders at the dispatch points in the production process and for issuing work to a workcentre. The implementation of a dispatcher may be automatic, in which case there is no human intervention in the task, semi-automatic, where a human selects a job to dispatch from a computer application and manual, where a human makes the decision of what job should be processed next.

The Monitor

The Monitor in Production Activity Control for SMEs is responsible for

1. Collecting data from the shop floor
2. Filtering the data such that only that required by the other elements in the system is procressed
3. Making the data available to the workload controller, the scheduler and the dispatcher.
4. Comparing the monitored data with the planned data.

Data collection from the shop floor will record the actual start times for each order on each operation. If the data collection records too much data for the scheduler then this data must be filtered. For example, if the start and stop time of every item in a batch is recorded and the schedule only requires a start time for the batch, then the excess data can be ignored by the monitor.

The filtered data must be made available to the other elements in the system. The means by which the data is transmitted may be manual or automatic and may be in batch or in real time. For example, it may only be required that the data be made available to the scheduler each time the schedule is being generated while the dispatcher may require real time data in order to make decisions which allow it to respond to disturbances and unexpected events.

One important aspect of the monitor is that it should be possible to compare the planned schedule with the actual achievements on the shop floor. This will help the human scheduler to identify the accuracy of the scheduler while also evaluating the performance of the shop floor.

The MIA

To control the shop floor in an SME accurate and timely information is required to support the production personnel. Although there is often an abundance of data available, this may not be in the correct format and may not highlight the important aspects of production to the relevant people.

The Manufacturing Impact Analyser (MIA) seeks to satisfy the need for relevant, accurate and timely shop floor information to everyone in the SME who requires it. This information may be of three types; historical, current or predictive.

2. Reactive Scheduling within Production Activity Control

In the past the tendency was to pursue the development of an *optimal* solution for manufacturing scheduling problems. However, systems developed in this manner do not have the sophistication necessary to deal with everyday occurrences on the shop floor. The main problem lies in the failure of solutions developed in research and development facilities to bridge the gap between the theory and the practice of scheduling. This view is supported by Duggan [3] who advocates that, "the scheduling problem as it is posed in the literature, does not exist in real factories". Therefore, instead of relying on the optimal type solution, the process of deriving any schedule should take account of the knowledge and experience that is available from shop floor personnel. It is essential that human intelligence becomes an integral part of scheduling strategies, as significant improvements can be made to heuristically based schedules by an intelligent user, who has intimate knowledge of the manufacturing system. The task of the human scheduler now becomes more than a simple task of planning and implementing a production schedule for a given scheduling period. A first pass schedule is generated using existing algorithms and heuristics and the human scheduler is now more concerned with adapting to changes in the production environment and reacting to unexpected events. Problems, such as varying process times, unreliable machines, lack of adequate capacity and faulty materials can mean that the scheduler must alter the initial schedule several times before the end of a scheduling period. In any shop floor control system therefore, there must be a facility for reacting to these disturbances.

Within the architecture defined by PAC, the human scheduler may react to unexpected events in two ways:

1. by altering the dispatch sequence or
2. by re-scheduling.

In order to be able to do this the tools must be available which allow interactive altering of both the dispatch sequence and schedule.

2.1 Dispatcher

Because of the limitations existing in SMEs we propose that the dispatcher for SMEs should be either semi-automatic or manual. Using a semi-automatic dispatcher a computer application will read the work-in-process and the schedule and provide the user with a list of jobs that should be processed. According to the schedule, the user will be advised to process a certain job. However, if it is required that this dispatch sequence be changed the operator will be able to do so by using the application to select an alternative. Using a manual dispatcher, the operator may have a dispatch list for a certain period, for example a day. The jobs on the dispatch list will again be prioritised according to the schedule, however if the operator wants to alter the sequence it is easily done.

Altering the dispatch sequence can be dangerous as the operator will not know the effects of the change on subsequent jobs. Problems with this could be minimised by shortening the dispatch period or by using the dispatch application to make changing of the default dispatch sequence difficult.

2.2 Interactive Scheduler

One of the most influential occurrences in Production planning systems in recent years has been the advent of the German Leitstand. The Leitstand is being used widely as a mechanism for distributed shop floor control in Germany [6]. A Leitstand has five major components [1]:

- A graphics component capable of representing the schedule in Gantt chart format
- A schedule editor which allows the user to change certain aspects of the schedule.
- A database management system which controls the information required by the scheduler.
- An evaluation component which lets the user compare the schedule which requirements
- An automation component for automatic generation of schedules.

The graphical component means that the user will not be confronted with technical skill requirements that differ greatly from those in use at present. It makes the scheduling process more intuitive for the user while also making it very easy for the scheduler to alter the schedule if required. When changing one job using the interactive scheduler, the user will see the effect of this change on all the subsequent jobs.

The interactive scheduler [5] is a tool which merges the graphical component and the schedule editor from the leitstand. Using this the schedule can be altered by moving blocks on an interactive display so that the knowledge of the human scheduler may be imposed on the scheduling process and to allow the schedule to change as disturbances affect the production system.

The Gantt chart representation is also used to display job history so that the human scheduler can very easily display the actual situation and compare this with the scheduled situation in order to assess performance.

3. The Manufacturing Impact Analyser (MIA): Decision support within PAC

The main purpose of the MIA in the architecture is to provide decision support to the human scheduler. To do this *predictive, what-if* and *information filtering* capability is provided.

3.1 Predictive Capability

The predictive capability extrapolates the behavior of the system from the current status to a point in the future. Using statistical data to model the uncertain behavior of the system each order may be assigned a *probability of lateness*. If the probability of lateness is greater than a pre-defined threshold then the user may be notified.

3.2 What if Capability

When making a decision the human scheduler wants to know how the change affects the job itself, but also how the change will affect the other jobs in the system both now and in the future. The what if capability of the MIA allows the user to experiment with changes to the system without altering the system itself. The user can then compare the experimental results with the original schedule and see if the changes were beneficial or not. If the changes are not beneficial or feasible then the changes may be discarded.

3.3 Information Filtering and Sourcing

Different people within an SME require different information about the production system. The production supervisor may want to know where a particular order is within the system, while the managing director may only want to know the success rate at meeting delivery dates. The abundance of data available must be filtered in order to meet the requirements of the person using the data.

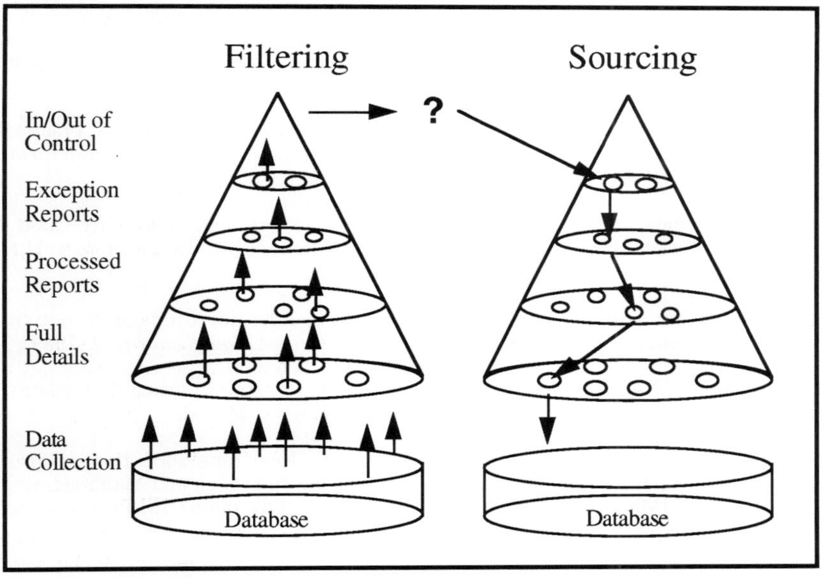

Figure 3: Information filtering and sourcing.

It is not merely enough to filter the data. If the person using the data requires to know the source of that data then they are allowed to move lower in the information filter to source the required data. The MIA exhibits this ability to source information so that when information at a higher level requires increased detail, the user is allowed to reverse the filtering process and is guided down through the levels to the source of the higher level information.

4. Conclusions

The task of controlling a shop floor in a SME requires the ability to react to unexpected events. Within an architecture for Production Activity Control the ability to react to unexpected events is provided by the use of a dispatcher and an interactive scheduler. Decision support for the scheduling task is achieved using the Manufacturing Impact Analyser.

A future area of interest is in the development of a knowledge based tool for storing the informal knowledge used in developing a schedule. The main thrust behind this is that a manufacturing enterprise should not lose all informal scheduling knowledge if the human scheduler decides to leave. This tool would behave as a decision support tool for assisting the scheduler while altering the schedule using the interactive Gannt chart.

References:

1. Adelsberger, H., and Kanet, J., "The Leitstand - a New Tool for Computer Integrated Manufacturing", *Production and Inventory Management Journal* - first quarter (1991): 43 - 48.

2. Bertrand, J. W. M., Wortmann, J. C., Wijngaard, J., " Production Control, A Structural and Design Oriented Approach.", *Elsevier Science Publishing Company Inc.* (1990)

3. Duggan, James and Browne J., "Production Activity Control: A Practical Approach to Scheduling" *International Journal of Flexible Manufacturing Systems*, 4 (1991):79 -103.

4. Higgins, Paul and Browne, J., "The Monitor in Production Activity Control Systems", Production Planning and Control, Vol. 1, No. 1, Taylor Francis, U.K. (1989).

5. Jackson, S., Browne J., "An Interactive Scheduler for Production Activity Control", *International Journal of Computer Integrated Manufacturing* 2 (1989).

6. Kohler, C., "New technological and organisational solutions to shop floor control" *Computer Integrated Manufacturing Systems*, 6 (1993) 44 - 52.

7. Mendonça J.M., J.W. Schulte, S. McCarthy, "Real time shop floor control: Modular system and integration tools", Advances in Design and Manufacturing: Realising CIM's Industrial Potential - Proceedings of the ninth CIM-Europe annual conference, 1993.

8. Anon. "This is Eolas" The Irish Science and Technology Agency, 1992.

Knowledge-Based Reactive Scheduling (B-15)
E. Szelke and R.M. Kerr (Editors)
Elsevier Science B.V. (North-Holland)

Reactive Scheduling in a Fuzzy-Temporal Framework

Jürgen Dorn, Roger Kerr* and Gabi Thalhammer

Christian Doppler Laboratory for Expert Systems
Technical University Vienna
Paniglgasse 16, A-1040 Vienna

Abstract

Practical scheduling usually has to react to many unpredictable events and uncertainties in the production environment. Although often possible in theory, it is undesirable to reschedule from scratch in such cases. Since the surrounding organisation will be prepared for the predicted schedule it is important to change only those features of the schedule that are necessary.

We show how on one side fuzzy logic can be used to support the construction of schedules that are robust with respect to changes due to certain types of event. On the other side we show how a reaction can be restricted to a small environment by means of fuzzy constraints and a repair-based problem-solving strategy.

We demonstrate the proposed representation and problem-solving method by introducing a scheduling application in a steelmaking plant. We construct a preliminary schedule by taking into account only the most likely duration of operations. This schedule is "repaired" until some threshold evaluation is found. The repair is basically a hill climbing search that tries to improve the evaluation of the schedule by repairing constraint violations. Finally, we show which events can lead to reactive scheduling and how this is supported by the repair strategy.

Keyword Codes: I.2.1;I.2.4;I.2.8

Keywords: Artificial Intelligence, Applications and Expert Systems; Knowledge Representation; Problem Solving

---

* University of New South Wales, School of Mechanical and Manufacturing Engineering, Kensington, New South Wales, Australia, recently Guest Professor at the Technical University Vienna

1 Introduction

Scheduling is a hard problem both in theory and practice. Theoretical scheduling problems, which are concerned with searching for optimal schedules subject to a limited number of constraints, suffer from excessive combinatorial complexity and are mostly NP Hard. Practical problems although more highly constrained, are complex due to the number and variety of the constraints themselves, many of which are "soft" e.g. potentially relaxable human preference constraints, rather than "hard" physical constraints. In addition an effective schedule often needs to be evaluated against a number of potentially conflicting goals which themselves may not be precisely defined, and many of the scheduling parameters such as processing times, material arrival times, resource availability etc. are also subject to uncertainty. Furthermore, in the dynamic environment of the shop floor, a variety of unexpected events are continually occurring and any schedule must in practice be subject to frequent revision to ensure it is in line with changing shop floor status. Scheduling is thus an ongoing and continuous process. The problem of updating schedules in the most effective way when the constraints or assumptions on which they are based are changed or invalidated is one that is receiving increasing attention amongst both researchers and practitioners, and is generally termed "reactive scheduling".

The main alternatives to the revision of a schedule in the presence of real time shop floor feedback are either to incorporate the new information by completely regenerating the original schedule from scratch, or by "repairing" the previous schedule in some way. The first approach might in principle be better capable of maintaining optimal solutions, but as pointed out above, such solutions are rarely achievable in practice, and computation times are likely to be prohibitive. Furthermore, frequent schedule regenerations can result in instability and lack of continuity in detailed shop floor plans, resulting in increased costs attributable to what has been termed "shop floor nervousness". Thus most approaches to reactive scheduling are based on infrequent regenerations of a basic predictive schedule which then maintains continuity by serving as a nominal reference for the identification and specification of schedule changes as it is progressively modified. Predictive and reactive scheduling may thus be seen as complementary activities (Smith 1993).

An important issue in which this complementary relationship between predictive and reactive scheduling is highlighted is that of schedule robustness. A robust predictive schedule is one which is likely to remain valid under a wide variety of different types of disturbance. Robustness is clearly a desirable attribute of a predictive schedule as it will reduce the number of subsequent reactive scheduling decisions required as the schedule is executed.

Robustness can generally be increased by the avoidance of unnecessary computational reaction to disturbances in a predictive schedule that have been specified to unrealistic degrees of precision. Although least commitment strategies are potentially useful in this respect, an alternative is to generate schedules in which the presence of uncertainty is recognised and explicitly represented. Temporal propagation of the uncertainty allows the possibility of the effects of unexpected events being localised to that part of the schedule where it is most certain they will have a significant impact rather than being propagated to far regions of the planning horizon where in reality they would be swamped by cumulative uncertainty arriving from other sources (Kerr and Walker 1989). Furthermore, evaluation of schedules incorporating uncertain processing times can provide a framework in which situations where critical events are tightly scheduled with little slack to be given a lower robustness score than those where slack exists.

In this paper we examine an application of fuzzy temporal reasoning to represent and propagate schedule uncertainty within the context of scheduling the operations of a steel-making plant. This will be seen to assist in the generation of robust schedules which result in considerable reduction in the amount of reactive scheduling required. The examples are taken from an application at the Böhler company, one of the most important European producers of high-grade steel. An expert system that supports the technical staff in the steelmaking plant in generating schedules for steel heats for one week has been previously developed (Dorn & Shams 1991). However the production rule based nature of this system only allowed shallow modelling of the expert's knowledge, and several enhancements have been proposed to generalize the problem-solving method and to use deep knowledge instead of shallow rules (Dorn 1993) and (Dorn and Slany 1993). These enhancements include explicit representation of uncertainty and vagueness in scheduling parameters using fuzzy sets.

2 The Steelmaking Process

The steel plant under consideration produces high-grade steel from crude steel using an alloying process for subsequent delivery to other plants which are concerned with forging and rolling. The basic processes to be scheduled are shown in block form in Fig 1. Crude steel is first melted with the relevant alloying metals such as manganese, tungsten, chromium, or others in an electric arc furnace (EAF). The melted steel is then processed in the secondary metallurgy (SM) first by pouring into ladles that are transported by a crane to a ladle furnace where fine alloying takes place, then on to a special treatment in a vacuum oxygen decarburation unit. The next step is the processing of the steel in a continuous caster (CC) to form slabs or the casting of it into moulds to form ingots. For jobs using the CC, if two sequential jobs using the same format are processed, the second job must be delivered in time to the caster to allow it to cast continuously otherwise a set-up time must be considered.

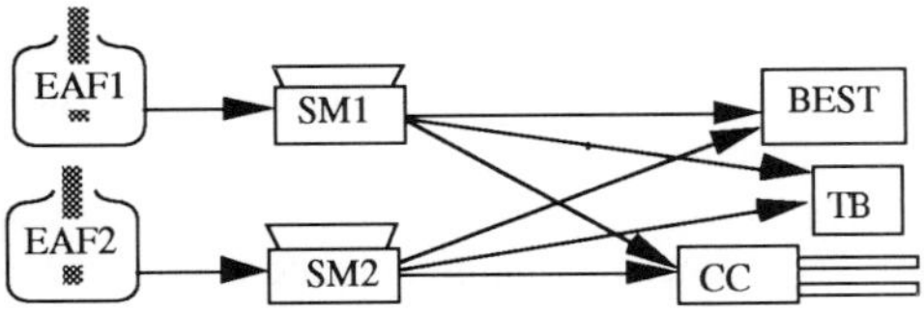

Figure 1: Simplified production lines of the plant

For solidification of ingots, the moulds require space in a teemimg bay (TB). For some large ingots, the Böhler-Electro-Slag-Topping (BEST)-technology (a special casting technology for ingots) is used. One space in the teemimg bay is dedicated to such ingots. The steel produced is then delivered to other plants including different rolling mills and forges which order slabs or ingots of a certain quality. Sometimes products are stored for several days in intermediate stock yards, because the plant that has ordered the product cannot process the jobs in the same sequence as the steelmaking plant. The differing sequencing criteria of jobs cause considerable costs for the company since a large amount of capital is invested in these semi-finished goods. Moreover, since the steel cools down it must be reheated in the next plant. To reduce costs and to improve the quality, some orders therefore have due dates at the subsequent plant which must be met as closely as possible.

As indicated in Fig 1, two production lines share the CC and the teeming bays. For every steel quality there is a process plan that prescribes on which line the job will be produced and which processes it will require. Once a week, engineers of the different plants meet to discuss orders for the next week. Compatible orders that may have different destinations are combined to form jobs. For each production line of the steelmaking plant a list of jobs for one week is worked out manually. The task of the scheduler is to find a possible sequence for all jobs which violate as few compatibility constraints as possible and to allocate resources over time without violating temporal and capacity constraints. The result of this scheduling process may be that some orders are rejected and shifted to the next week. One objective for the development of the expert system was to reduce the number of rejected orders.

2.1 Constraints in Scheduling

During the construction of a schedule several constraints have to be considered. These constraints are often vague and conflicting. The engineer has no pretension to generate an optimal schedule because he knows that the uncertainty in the execution of his plan would break this optimality, but he does know that some schedules are better than others. Three kinds of constraints may be distinguished: compatibility, temporal, and capacity constraints. Compatibility constraints between the permissible relative chemical compositions of sequentially processed jobs are the most important constraints on the schedule. Residuals of one heat in the electric arc furnace may pollute the next heat and so the engineers use as a rule of thumb that 3 % of a chemical element in a heat remain in the wall of the electric arc furnace and 3 % of the difference of the elements in two consecutive heats will be assimilated by the second heat. Of course, the 3 % are always on the safe side and in some cases the expert relaxes this factor. However since these constraints are not central to the issue of reactive scheduling, we do not consider them further here and refer to (Dorn and Slany 1993) for more details.

Since some steel qualities require the cast steel to be hot for the subsequent treatment like forging, due dates exist that should preferably be met to within a tolerance of ± 2 hours. The average number of jobs with such due dates is about 10 %. These due date constraints are not really hard since they may be relaxed through negotiations with the subsequent plant.

Waiting time between operations means either that the steel cools down and must be reheated or it must be heated continuously during waiting. Both cause costs and for some steel grades, unfavourable effects on the quality may result. The scheduler must therefore guarantee that slack times between the successive operations of a job may not exceed a certain limit. An objective for production is thus to have slack intervals as small as possible, resulting in minimization of the flow time and reduced production costs. Also, jobs on the CC should be scheduled either continuously or with sufficiently large breaks to allow set-up and maintenance. Since the CC is a resource which is shared between the two production lines, it is important to schedule it effectively and to ensure that jobs arrive at the correct time.

BEST-ingots introduce some problems. Typically there are groups of such jobs which tend to be low alloyed, forging grade ingots with the same chemical quality requirements. From the compatibility view point they should be produced in sequence. However since only one place exists for them in the teeming bay, and the solidification time tends to be considerably longer than the other process times, they cannot be scheduled immediately after one another. This can present a problem especially for big ingots which have a very long solidification time. Typically they need to be scheduled with sufficient gap between them to allow for solidification of one job in the teeming bay to be complete before the next job arrives.

2.2 Uncertain Process Times and Reactive Scheduling

Individual processing times of jobs depend on the type and quantity of steel being processed. The duration of the secondary metallurgy (SM) and the casting (CC) processes are to some extent controllable so that slack time can be reduced. This controllability will be ignored in the discussion that follows which centres on the uncontrollable and unpredictable components of the process time only. All processes have an inescapable element of variability, which may be captured in the form of possibility distributions. Assuming jobs of average size, the following diagrams represent estimates by experienced schedulers of the possibility distributions of the times required for different processes. Each diagram represents the degree of membership of the natural numbers within the indicated range, of the fuzzy set of time durations for that process. Time durations are thus measured by trapezoidal fuzzy numbers. However, for the purpose of improving the understandability of temporal propagation in subsequent discussion we also give three characteristic durations, i.e. the minimal, the most likely and the maximal duration. The most likely duration can be computed from the fuzzy sets by determination of the center of gravity.

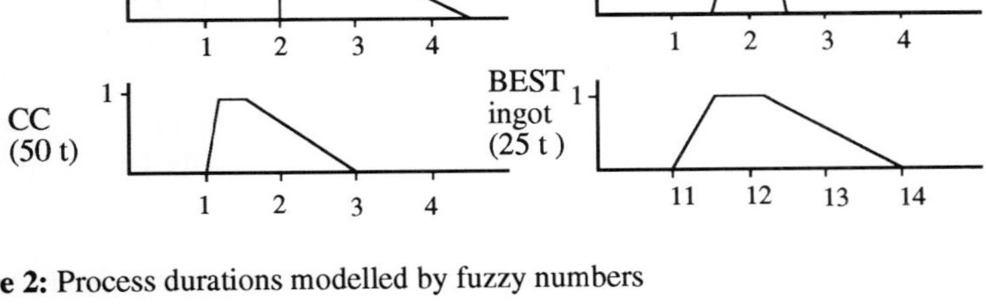

Figure 2: Process durations modelled by fuzzy numbers

Table 1: Durations of operations

duration \ aggregate	EAF	SM	CC	BEST
minimal	2	1.5	1	11
most likely	2.5	2	1.5	12
maximal	4.5	2.5	3	14

From the extent of the variability shown by these process times, it is clear that potential problems can occur as a result of jobs taking longer than their expected process times. A schedule that has been constructed to satisfy the relevant constraints can be invalidated due to constraint violations when jobs take other than the process times assumed for the purposes of constructing the predictive schedule. Since the bottleneck resources tend to be the Electric Arc Furnaces which are run continuously, a single job taking longer than normal to process on the EAF can result in all subsequent jobs being later than their scheduled times, thus causing due dates to be missed. This can also have serious implications for jobs processed on the CC. For example if a job arrives from one production line for processing on the CC but this is still processing a late job from the other production line, the first job may be destroyed if it has to wait too long. The same type of problem may occur with jobs using the BEST technology where if the dedicated teeming bay is still occupied by the previous BEST job which is late, the job requiring the teeming bay may be destroyed.

In practice, the engineers who schedule the plant attempt to construct schedules that are reasonably robust with respect to variabilities in the process time. Thus jobs from different production lines are scheduled on the CC with gaps between them to try and minimise the possibility of conflicts resulting from lateness. BEST jobs are also scheduled at a maximum of two per day to minimise the possibility of more than one BEST job requiring space in the teeming bay. The predictive schedule is prepared for a period of one week ahead, and during the course of its execution, the engineers make judgmental decisions as to whether or not it will require modification in the light of unfolding events. For example a job that takes longer than usual to process on the Electric Arc Furnace may, in a crisp evaluation of the schedule, result in a subsequent job being predicted to just miss its due date. However, rather than rescheduling this job, the engineer knows that the degree of uncertainty in the events still to occur before that job is processed will mean that this minor constraint violation is not yet significant at this stage and will not perform any rescheduling.

An Expert Scheduling System was constructed for use in the plant (Dorn and Slany 1993) which used crisply defined average process times for the construction and evaluation of the schedule. This was based on the use of simple heuristics similar to those used by the human scheduler to construct an initial schedule which was then evaluated against the applicable constraints, and subject to a "repair" strategy to attempt to improve it by reducing the number of constraint violations. This technique is clearly directly applicable also to the reactive scheduling problem in which unfolding events generate constraint violations which in principle could be subject to repair using the same strategy as the predictive scheduler. However the assumption of crisp process times made in the original scheduler did not allow the possibility of grading reactively generated constraint violations according to the degree of uncertainty surrounding them, and of using this as a basis for selectively repairing only those violations that were significant. Furthermore, use of crisp process times also inhibited the evaluation of schedules according to their relative degrees of robustness against contingencies, and the establishment of trade-offs between robustness and cost (which the human schedulers performed implicitly). In the next section we describe, in the form of a simple example, a modified form of the predictive scheduler which uses fuzzy rather than crisp process times for the evaluation of schedules in order to partially overcome the problems outlined above.

3 Schedule Generation

The strategy used by the system to generate a schedule is to use constructive heuristics based on those used by human schedulers to produce an initial schedule assuming that all processing times take on their most likely values. Jobs are scheduled in order of their criticality (defined later) with no backtracking. Some constraints may hence be violated. However when the schedule is evaluated for such constraint violations, the crisp processing times are replaced by their fuzzified possibility distributions, and temporal events are predictively propagated using rules for the addition of fuzzy numbers. This results in temporally distant events whose timing has been computed by the addition of several fuzzy processing times having broader possibility distributions than temporally close events whose timing is dependent on the execution of only one or two processes.

Temporal constraint violations are thus not expressed in a binary satisfied/unsatisfied form but as a matter of degree, with each temporal constraint having a degree of satisfaction in the range [0,1] based on the amount of overlap of the temporal estimates of events whose relationships are constrained. Thus the extent to which a due date constraint is met is measured by the degree of overlap between the due date (which may itself be fuzzily defined) and the fuzzy estimate of the finishing time of the job.

Threshold values, which may be zero, can be defined for acceptable and unacceptable degrees of constraint satisfaction (which can vary according to the type of constraint). If the degree of satisfaction of a constraint is unacceptable, then the schedule must be repaired by rescheduling in such a way that the threshold value for the constraint is met. The evaluation of the total schedule is performed by combining the degree of satisfaction of the individual constraints (which may themselves be weighted) in some appropriate way. A simple hill climbing technique is used to attempt to improve the schedule score by progressive interchange of jobs.

This also allows the incorporation of robustness into schedule evaluation. For example it is critical that jobs scheduled on the CC which originate from different production lines do not overlap. However, the closer they are scheduled together on the CC, the higher will be the degree of overlap of the fuzzy estimates of the time period over which they will require the CC, and the lower will be the robustness of the schedule. This fact can be incorporated into the schedule evaluation function with a weighting factor that gives the appropriate trade-off between robustness and cost. If a job takes longer than its normal processing time but (for example) this does not result in any due date constraint satisfactions dropping below its critical threshold, no rescheduling action need necessarily be taken other than the routine application of the hill climbing improvement technique.

By means of a small demonstration we show how a schedule is constructed, evaluated and repaired when presented with unfolding events. We assume that within the planning horizon a preliminary schedule can be built from the given jobs for each production line. We start with one production line and later extend this example to the second line to highlight the concept of schedule robustness.

3.1 Generating a preliminary schedule with domain heuristics

For simplicity, we restrict ourselves to a planning horizon of one day and we assume that we can schedule eight jobs. The preliminary schedule is constructed with simple heuristics which only implicitly consider the different constraints. The duration of the operations are represented crisply with a relatively large granulation. We take two and a half hour melting time in the arc furnace, two hours in the secondary metallurgy, one and a half hours on the caster and five and a half hours for normal ingots, and twelve hours for the BEST-ingots.

The jobs to be scheduled are displayed below in Table 2. The job_no identifies the job (e.g. j_3^1 means first production line, third job). The column type reveals which product, a slab or an ingot with or without BEST-technology, should be processed. The due_date imposes an additional time constraint on some jobs. The due date should be met by the finishing time of the last operation of the job. The importance of jobs is given a priori and is related to customer/market considerations. The table shows linguistic variables for the importance. From the importance of jobs and their specific characteristics a criticality value is computed that is used to decide in which sequence the jobs will be scheduled. An example for such a fuzzy rule is as follows:

IF the proportion of BEST-jobs is *greater* than 0.3 **AND** a job J is a BEST-job **AND** the *importance* of a job J is *high* **THEN** the *criticality* of job *J* is *very high*

Table 2: List of jobs to be scheduled on the first production line

job_no	type	due_date	importance	criticality
j_1^1	BEST		*medium*	*very high*
j_2^1	BEST		*medium*	*very high*
j_3^1	slabs	6 am	*very high*	*high*
j_4^1	ingots	3 pm	*high*	*medium*
j_5^1	slabs		*medium*	*low*
j_6^1	ingots		*medium*	*very low*
j_7^1	ingots		*low*	*very low*
j_8^1	ingots		*very low*	*very low*

As there is only one place for BEST-ingots which need a long solidification time compared to the preceding processes, the jobs that produce BEST-ingots may not be scheduled immediately after another. An interval should be between these jobs that allows the first ingot to have solidified by the time the second is cast. In this case the heuristic assumes an interval of twelve hours should be left between BEST jobs (which can of course be later relaxed to some degree according to prevailing circumstances). The jobs processed on the caster should be scheduled in sequence so that the two heats can be cast continuously. Jobs with a due date will be scheduled next. As there is always enough room in the teeming bay for normal ingots no capacity constraint is posed there. The system now generates a schedule by scheduling the important or most difficult jobs as described as follows:

First we schedule the two "BEST"-jobs (j_1^1, j_2^1) as far apart as possible. The first starts at 2 pm which shall be the start of our planning horizon and the second is scheduled at the end. Since we assume that each job takes two and half hours on the electric arc furnace and we have eight jobs, the second "BEST"-job is scheduled 7.30 am.

Next we schedule the caster job with a due date (j_3^1). We back schedule the operations of this job from the time due, assuming crisp processing times, and obtain a scheduled start time of 12.00 pm for this job. The second job with a due date is j_4^1. Again, scheduling backwards from the time due, and assuming the solidification time is five and a half hours, a starting time of 5 am is assigned to this job. Now the second job that is processed on the caster can be scheduled in sequence with the other "caster"-job. Finally, we also assign start times to the three remaining jobs in order of their criticality value and achieve the schedule in the following figure.

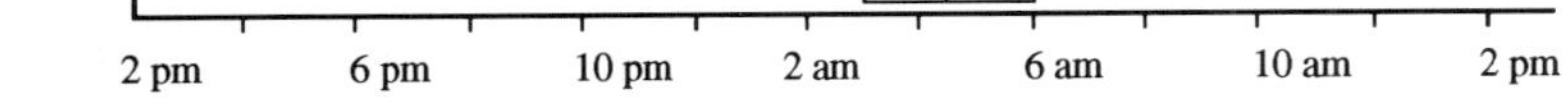

Figure 3: Preliminary schedule for first production line

3.2 Schedule Evaluation

A schedule is now evaluated in terms of the degree of satisfaction of the applicable constraints. Schedules with high satisfaction scores are good schedules. A schedule for which the degree of satisfaction of any constraint is below the critical threshold for that constraint is not feasible. Since it is not always possible to find a feasible schedule for all jobs, we also evaluate the schedule in terms of the number of jobs successfully scheduled, weighted by their average importance.

We assume the importance of each job is given by a fuzzy value denoted by the function *importance(J_i)*. Further, we assume that n is the number of existing jobs. We compute the mean importance of all jobs that are scheduled. In our example we achieve a mean importance that is medium. The degree of the satisfaction of an involved constraint is given by the function *satisfaction(C_j)*. Since not all constraints have the same importance for the schedule evaluation, the satisfaction is weighted by a type-specific factor. Although a constraint satisfaction which is below the critical threshold (which can also represent violation of a hard constraint) should lead to a refusal of a schedule, in the evaluation function a strongly violated constraint which is nevertheless above its threshold may be neutralized by many unviolated constraints. The evaluation of a schedule S is then:

$$f(S) = \frac{1}{n} \sum_{i=1}^{n} J_i \in S \wedge importance(J_i) + \sum_{j=1}^{m} C_j \in S \wedge satisfaction(C_j) * weight\ (type(C_j))$$

As mentioned previously, a very important set of constraints deals with chemical compatibility between adjacent jobs. However to give better visibility to the temporal aspects of the reactive scheduling problem, we shall not consider them explicitly here. We thus have three explicit constraints in our problem. We have two due dates for the jobs j_3^1 and j_4^1 and the constraint that j_1^1 and j_2^1 must be separated to the extent that the possibility they will conflict in their demand for space in the teeming bay is acceptably low.

To model the evaluation of the due date constraint, we define a fuzzy set A which is the set of all jobs that approximately meet their due date dd within the interval of [dd-2, dd+2] hours. The membership function will evaluate to $m_A(x) = 1$ if the due date x is met within the interval of [dd - 1, dd + 1] hours which is a *in time* value for this constraint. It will evaluate to $m_A(x) = 0.6$ for a due date violation of ± 2 hour which is still *early / late*. And it will evaluate to $m_A(x) = 0.3$ for a violation of ± 3 hours which is just *very early / very late*. Outside the fuzzy set A it will evaluate to $m_A(x) = 0$ for a violation larger then ± 4 hours. The membership function values in between can be linguistically described in terms of e.g. almost very good for 0.7 and so on.

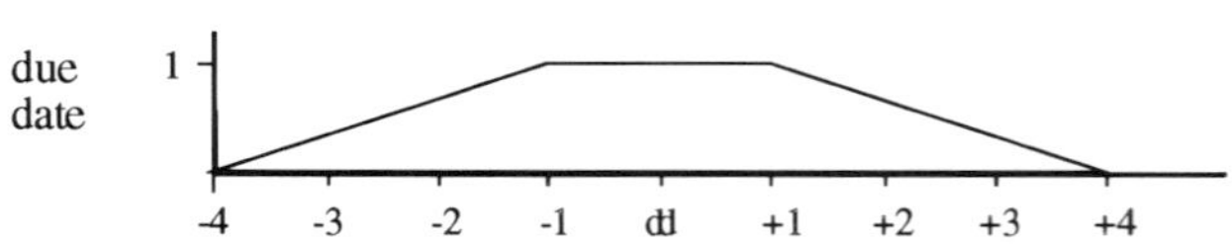

Figure 4: The fuzzy function for the evaluation of the due date constraint

For our preliminary schedule we now have to propagate forward from the crisply defined start of the planning horizon to obtain a fuzzy estimate of the predicted completion times of the jobs with due dates. This is obtained by addition of the fuzzy process times defined in Fig 2 for all jobs that must be processed on the EAF prior to the jobs with due dates, (which determine the fuzzy start times of these two jobs) and the fuzzy process times of the jobs themselves. Addition is performed by rules for the addition of the fuzzy numbers defined in Fig 2. The boundaries on the resulting fuzzy finishing times for j_3^1 are shown in Fig 5 which displays the propagation of earliest start and latest finish times of jobs up to and including j_3^1. The resulting fuzzy finishing time of j_3^1 is shown in Fig 6 in relation to the fuzzy due date.

As four jobs are preceding j_3^1 and each adds about 2 hours possible delay on the EAF the starting of j_3^1 could be delayed by 8 hours at the most (see table 1). The fuzzy duration of j_3^1 shows that it can take +2 hours on the EAF, +0.5 hours on the SM and +1.5 hours on the CC. So the total latest finish time possibly predictable will be the scheduled finish time which is 6 am + 8 hours + 4 hours = 6 pm (which is almost impossible as expressed by the low membership degree in Fig 6). Even though the operation on the SM of the preceding jobs can also vary it will not change the total latest time as the worst duration on the EAF will dominate.

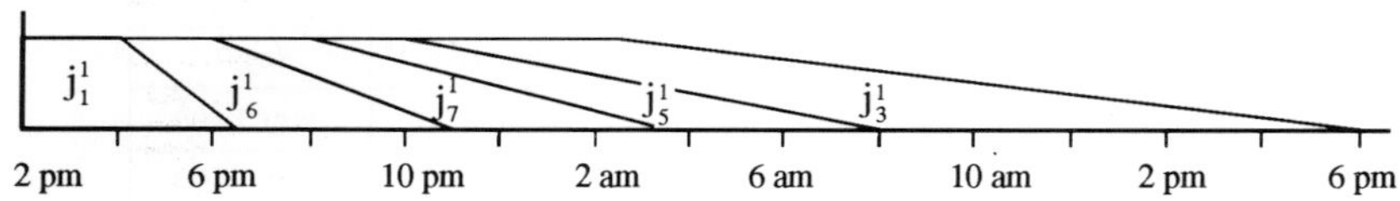

Figure 5: Fuzzy starting times of j_3^1 obtained by fuzzy temporal propagation

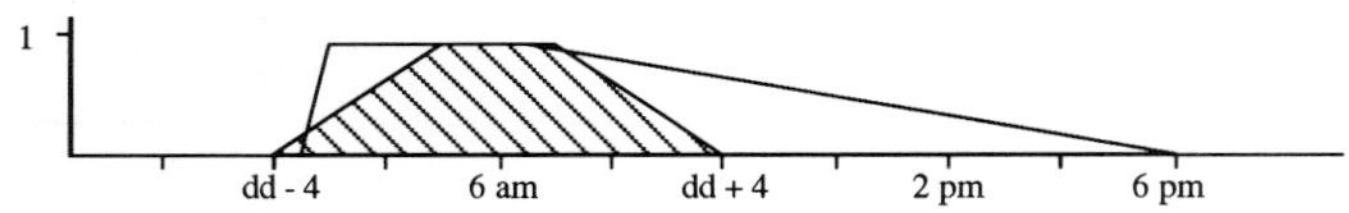

Figure 6: Degree of satisfying the due date constraint for j_3^1

To determine the satisfaction of a due date, we apply a set of rules that map the topological features of the overlap between the fuzzy finishing time and the fuzzy due date to the interval [0,1]. These values can in turn be mapped to linguistic variables. We assume that our rules give a degree of satisfaction of 0.8 for the job j_3^1. If we assume that the critical satisfaction threshold for the due date constraint is set at 0.4 a further set of rules could map this degree of constraint satisfaction to the linguistic variable "high". As seen in Fig 6 the overlap within the range of the most likely duration is very large.

Fuzzy temporal propagation is also used to determine whether there is enough time for the first BEST-ingot to solidify before the second one is cast. The overlaps of the fuzzy estimates of the finish time of the first BEST job and the start time of the second are evaluated by propagation. The rules which map the topological features of this overlap to the [0,1] interval generate a degree of satisfaction of 0.9 for this constraint, which is above the critical threshold of (say) 0.5 and is in turn again mapped to the linguistic variable "high" (ie. resulting in a high satisfaction of this constraint). The overall evaluation of the preliminary schedule in terms of constraint satisfaction can then be determined using rules which combine the degrees of satisfaction of the individual (weighted) constraints. We assume that these rules result in a score of 0.9 for the overall schedule, which may also be mapped to the linguistic variable "high".

3.3 Robustness of Schedules

We will now add the second production line to our example. Again we look at the list of jobs and try to build a schedule by first taking the important and difficult jobs into consideration. We will not explain how the normal ingots are scheduled and concentrate on the critical jobs which use the CC and the solidification bay for BEST-ingots. We have two jobs with due dates and two jobs that have to be scheduled on the caster as can be seen in the next table.

Table 3: List of jobs to be scheduled on the second production line

job_no	type	due_date	importance	criticality
j_1^2	ingots	7 am	*high*	*high*
j_2^2	ingots	1 pm	*medium*	*high*
j_3^2	slabs		*low*	*very high*
j_4^2	slabs		*low*	*very high*
j_5^2	ingots		*medium*	*low*
j_6^2	ingots		*low*	*very low*
j_7^2	ingots		*very low*	*very low*
j_8^2	ingots		*very low*	*very low*

Since we have also two CC-jobs on the other production line these jobs are critical in terms of attempting to avoid conflicts for the CC. They should be scheduled in sequence, however, the due dates for the other two jobs must also be met if possible. We also have to guarantee that even if the jobs of the second production line scheduled on the CC are late they will not conflict with the jobs of the first production line. So the further we schedule those jobs apart the more robust our schedule will be. We assume we have constructed with our heuristics the following schedule:

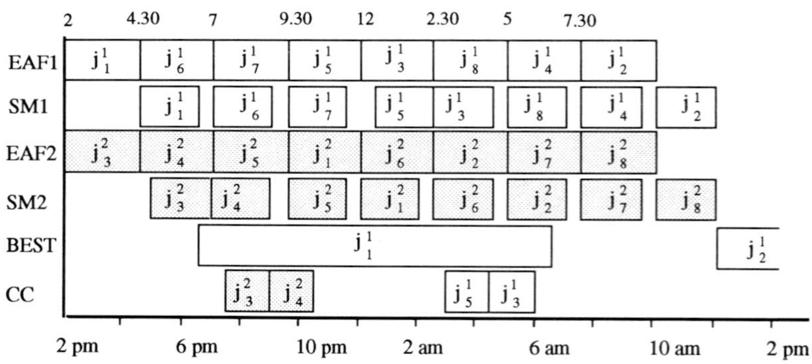

Figure 7: Preliminary schedule for both production lines

The predicted fuzzy finishing time of j_4^2 on the CC is shown in Fig 8 in relation to the predicted fuzzy start time on the CC of j_5^1. It will be seen that the presence of process time uncertainties leads to an overlap between these two events which implies the possibility of conflict. A set of rules is then invoked which reflects the human schedulers attitude to this degree of overlap in terms of schedule robustness, and we shall assume that it maps to a value of 0.3 in terms of the degree of robustness of the schedule with respect to this particular CC conflict constraint. If we further assume the mapping rules were developed on the basis of the acceptability criterion for degree of satisfaction of this particular robustness constraint being 0.5, then it is apparent that this schedule is unacceptable with respect to robustness, and must be repaired.

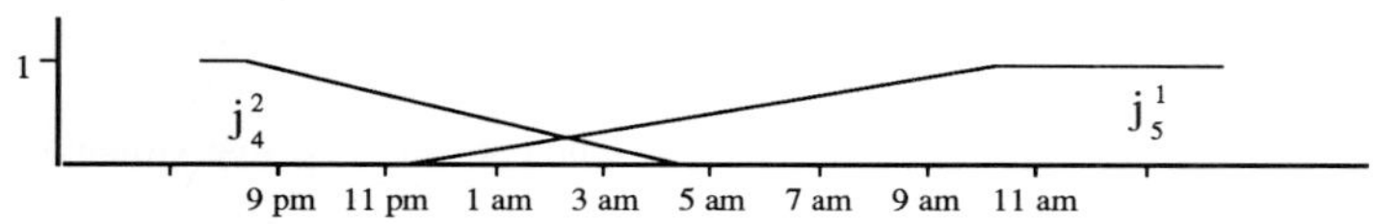

Figure 8: Potential conflict between jobs from different production lines that use the CC

3.4 Repair-based Strategy

The repair-based search strategy is based on trying to improve the overall schedule evaluation in terms of fuzzy values. It is a iterative process consisting of single repair steps. The procedure determines first a significant constraint violation (ie. a constraint satisfaction which is below its critical threshold) and tries to repair it. By applying the repair step to satisfy a violated constraint of our schedule, we achieve a new schedule whose evaluation function must be improved.

The repair process must iterate as long as there are any degrees of constraint satisfaction below their thresholds. Afterwards the repair is used to optimize the schedule further by attempting to increase the degree of satisfaction of remaining constraints that are above the threshold level. As a stopping criterion for the optimization we use the ratio between achieved improvement of a repair step and the number of applied repair steps. If this ratio becomes smaller than a predefined value we stop our search.

In its simplest form, a repair step is an exchange of two jobs that improves the schedule. For example, in the case of a due date constraint one job is involved. To repair an unacceptable constraint satisfaction value, the job is exchanged with any other job and the new schedule evaluated. If the constraint satisfaction degree has increased to above its threshold, and no other constraint satisfaction has dropped below the corresponding threshold then the repair step is finished. Otherwise, other exchanges have to be tried. Sometimes we have more jobs than we can schedule. In this case an exchange with a job in the list of unscheduled jobs is also allowed.

In the example discussed, the only unacceptably satisfied constraint relates to the robustness issue, and is that a conflict may occur on the caster. The repair-based algorithm tries to reschedule either j_4^2 or j_5^1. The exchange of the jobs j_5^1 and j_8^1 results in best evaluation of the overall schedule. In this case, no other constraint will be violated and the violation of the robustness constraint is repaired.

In general, a repair step can consist of several exchange operations in order to escape local maximas. Several search strategies seem to be appropriate to find a repair step consisting of single exchange operations. We have used a depth-first search with a depth bound set to five. The search is stopped if a schedule is found whose evaluation has improved and the constraint violation is repaired. Initial experiments have shown that we always find a good feasible schedule within a quite small number of repair steps.

4 Reactive Scheduling

Repairing a schedule during its first construction is a similar process to the reaction to unexpected events. Such an event may cause a constraint to be violated. For example, if one operation takes longer than expected, a due date constraint may be violated. We can now evaluate this violation and if necessary, apply the repair strategy to solve it. We assume that through the occurance of event *E* the evaluation of our schedule gets worsened. The optimal reaction would be that we find a set of repair steps so that the new evaluation would be as good or better than the old evaluation (before *E*).

However, to avoid frequent changes of the schedule due to very small deviations between predicted and actual times, we should define thresholds as to what events are significant. The first threshold is concerned with the event itself and is set so as to prevent re-evaluating the schedule as a result of events that are obviously trivial. Re-evaluation should only be performed if the event *E* is regarded as significant as measured for example by occurring at a time that deviates by an amount greater than some absolute threshold (eg 10 mins) from its predicted time. The second threshold relates to the degree of change in the schedule evaluation score, or in degrees of satisfaction of individual constraints, that can be tolerated before repair is considered necessary.

As a simple example, consider that the melting of job j_1^1 is finished in two hours rather than the expected 2.5 hours. If we do a new evaluation that consists of a new temporal propagation, the likelihood that the subsequent jobs will finish earlier than expected is greater, since they all start earlier. The fuzzy finish time of job j_3^1 can be shown in relation to its fuzzy due date. In Fig 9 job j_3^1 has a slightly changed latest finish time (compared with Fig 5 and Fig 6). For the latest finish time of j_3^1 we assume all prior jobs to take 2 hours longer, but now we can reduce this uncertainty as job j_1^1 has definitly taken 2 hours instead of possibly four and a half hours. So we are gaining two and a a half hours.

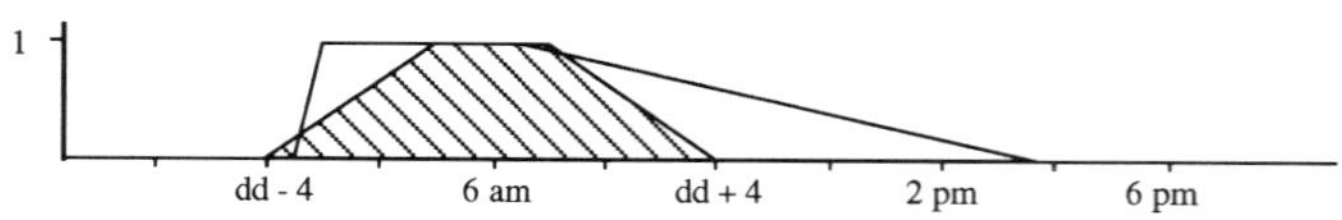

Figure 9: Degree of satisfying the due date constraint for j_3^1

The fuzzy finishing time does not extend so far down the time horizon since there are fewer uncertain process times remaining on which the timing of this event depends. However our rules for mapping of the topology of the overlap with the fuzzy due dates on to

[0,1] do not change the degree of constraint satisfaction, so no repair is necessary. This would constrast to the crisp case in which use of average times would lead to the prediction the job would be too early and would have to be rescheduled. As we progressively execute the schedule, the width of the fuzzy estimates of the finishing time of this job progressively narrows, to the point where it is sufficiently accurate as to tell us whether or not a significant problem exists and rescheduling should occur.

As a further example, consider the case in which melting of both jobs j_3^2 and j_6^1 are finished at 6 pm. In this case, the critical constraints are the degree of satisfaction of the robustness constraint relating to the caster, and the degree of satisfaction of the due date constraint for j_1^2, which is now more likely to be too late. The relevant fuzzy time estimates are again obtained by fuzzy propagation. The following figure shows the original evaluation of the robustness constraint in the dark plane and the robustness evaluation of the repaired schedule in the light plane.

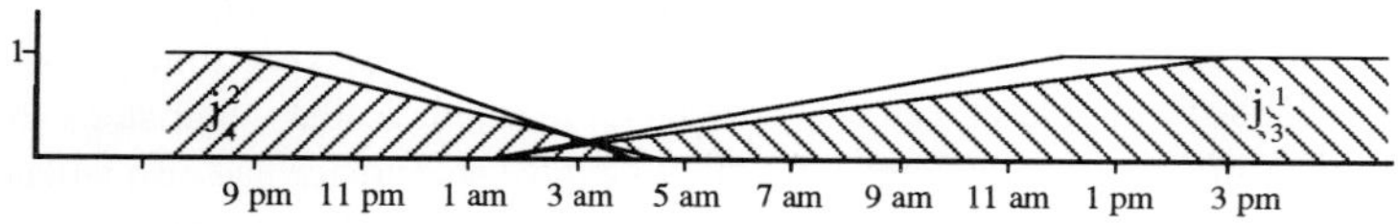

Figure 10: Potential conflict between jobs from different production lines that use the CC

Examination of the new overlap between the fuzzy finishing time of j_4^2 on the CC and the fuzzy start time of j_3^1, shows that this overlap has been slightly reduced due to the fact that some of the uncertainty in the schedule has been removed by the execution of three jobs (their process times being removed from a range of possibilities and placed in the category of crisp values). Thus although the actual events appear to have pushed the expected execution times of the two pairs of jobs requiring the CC closer together, the schedule itself is actually more robust than before because the event timings are known more precisely.

We find however that late finish of job j_3^2 results in a problem with j_1^2 now being too late (giving a due date constraint satisfaction function of below its threshold). Thus although no rescheduling is required from a robustness point of view, we do require to reschedule from the point of view of the due date constraint of j_1^2. As before, the iterative algorithm searches for steps to repair this constraint violation. The best choice is to exchange j_1^2 and j_5^2.

5 Conclusions

This paper has demonstrated by means of a simple example, how fuzzy temporal representation of schedules, coupled with a repair-based scheduling strategy, can assist in the explicit representation of the robustness of a predictive schedule, and through the setting of constraint satisfaction thresholds, in the evaluation of whether and the extent to which a predictive schedule needs to be modified during the progress of its execution. At present, the degree of robustness of the schedule is measured simply in terms of degrees of overlap of certain critical events. Other measures of robustness are obviously possible and work is in progress on the development of a separate expert system that evaluates robustness in a more sophisticated way. Clearly, the behaviour and hence the effectiveness of the system will also to a large extent depend on the threshold values of each constraint. These values therefore require a considerable degree of fine tuning in an environment in which some measure of effectiveness of schedule reactivity can be systematically related to individual threshold values. It is anticipated that simulation will be used to assist in this task. The scheduling system is also in the process of extension to represent additional temporal constraints introduced by shifts, the working hours of subsequent plants, and production set-ups and maintenance operations. The issue of controllability of process times as discussed in (Dubois *et al.* 1993) and the implications of this controllability for reactive scheduling and robustness are also being investigated.

The iterative repair algorithm is also still under consideration. Usually in iterative algorithms like Tabu-search (de Werra and Hertz 1989), MinConflicts strategy (Minton *et al.* 1990) or anytime rescheduling (Zweben *et al.* 1990) only local modifications are tried. We search for repair steps in the entire schedule. This can be very time consuming. Looking at the last example we can also see that more sophisticated repair steps are possible. Consider the case that the long duration of j_3^2 in the arc furnace had also violated the due date of j_2^2. The described iterative algorithm would solve both constraints separately in two repair steps. However, an exchange of the jobs j_1^2, j_6^2, and j_2^2 with job j_5^2 would be more appropriate. Experiments with case-based reasoning are performed at the moment to learn such repair steps.

6 References

de Werra, M.and Hertz, A. (1989) Tabu Search Techniques: A Tutorial and an Application to Neural Networks *OR Spektrum* **11**, pp. 131–141.

Dorn, J. and Shams, R. (1991) An Expert System for Scheduling in a Steelmaking Plant, Proceedings of the World Congress on Expert Systems, Orlando Fla., pp. 395–404.

Dorn, J. (1993) Supporting Scheduling with Temporal Logic, Proceedings of the IJCAI'93 Workshop on Knowledge-based Production Planning, Scheduling, and Control.

Dorn, J. and Slany, W. (1993) A Flow Shop with Compatibility Constraints in a Steelmaking Plant, in Drummond and Zweben (eds) Intelligent Scheduling, Morgan Kaufmann.

Dubois, D. Fargier, H. and Prade H. (1993) The use of fuzzy constraints in job-shop scheduling, Proceedings of the IJCAI'93 Workshop on Knowledge-based Production Planning, Scheduling, and Control.

Kerr, R.M. and Walker, R. N. (1989) A Job Shop Scheduling System Based on Fuzzy Arithmetic, Proceedings of the 2nd International Conference on Expert Systems and Leading Edge in Production and Operations Management, Hilton Head Island, S.C., pp. 433–450.

Minton, S.,Philipps, A., Johnston, M. and Laird, P. (1990) Solving Large Scale CSP and Scheduling Problems with a Heuristic Repair Method, Proceedings of the 8th National Conference on Artificial Intelligence, pp. 17–24.

Smith S. F. (1993). OPIS: A Methodology and Architecture for Reactive Scheduling; in Fox M., and Zweben M. (Eds) Intelligent Scheduling , (Morgan Kaufman Publishers)

Monte Zweben (1990) Anytime Rescheduling, Proceedings of the Workshop on Innovative Approaches to Planning, Scheduling and Control, pp. 251–262.

Knowledge-Based Reactive Scheduling (B-15)
E. Szelke and R.M. Kerr (Editors)
Elsevier Science B.V. (North-Holland)

HOW TO APPLY FUZZY LOGIC TO REACTIVE PRODUCTION SCHEDULING

Günter Schmidt

Universität des Saarlandes, Postfach 1150, D-66041 Saarbrücken, Germany

ABSTRACT
An approach modelling reactive scheduling by fuzzy logic is presented. The purpose is to get closer to human problem solving in situations where pure combinatorical approaches do not work any longer. Here rather intuition and expert knowledge is required. Applying fuzzy logic we do not solve reactive scheduling problems but we diagnose when steps for problem solving have to be undertaken.

Keyword Codes: I.2.1; I.2.4; I.2.8
Keywords: Artificial Intelligence, Applications and Expert Systems; Knowledge Representation; Problem Solving

1. INTRODUCTION

Within all activities of production planning and control reactive production scheduling is a major part of production management. Production scheduling means the task of resource allocation over time considering a discrete, distributed, dynamic and stochastic environment. It is carried out not only on a reactive or online control (ONC) level, but also on a predictive or offline planning (OFP) level. OFP generates the requirements for ONC and ONC gives feedback to OFP. Predictive scheduling serves to provide guidance in achieving global coherence in local decision making. Reactive scheduling is concerned with revising predictive schedules as unexpected events force changes.

With this paper we want to present some ideas how to model parts of the reactive scheduling process using fuzzy logic. The idea is to use this approach for monitoring and diagnosing purposes. Based on the corresponding observations detailed reactive scheduling actions can be taken. The paper is organized as follows. First we review quite generally existing solution approaches to scheduling problems. Then we present our approach to apply fuzzy logic to reactive production scheduling. Finally we discuss an architecture to combine predictive and reactive production scheduling.

2. SOLUTION APPROACHES TO SCHEDULING PROBLEMS

There are two major steps which have to be covered by production scheduling [Sch 89].

(1) System Operation: here job dispatching for entry into the manufacturing system, job traversing inside the system and job sequencing for processing have to be decided in accordance with business objectives and constraints.

(2) System Supervision and Monitoring: using process data the progress of job processing has to be checked continuously. The actual state of the manufacturing system should always be transparent, in order to be able to react quickly if deviations from a planned state are observed.

Reactive scheduling is related to both steps. The arising problems in organizing short term system operation are difficult to solve; there is not only the inherent combinatorial complexity of the scheduling problem but also that input data is voluminous and rapidly changing so that it is difficult to have the right estimate about the manufacturing situation. For example, new orders arrive, others are skipped or the availability of resources changes suddenly. The lack of stability makes it necessary to have permanent revisions based on monitoring and diagnosing the actual manufacturing situation. Solution approaches to be applied in such an environment should handle vague data and must have especially short computing times, i.e. time- and resource-consuming models and methods are not appropriate for reactive production scheduling.

All models and methods for production scheduling so far developed either have descriptive or constructive properties. Descriptive models give an answer to the question "what happens if ... ?"; constructive models try to answer the question "what has to happen that ... ?". Constructive models are used to find best possible or simply feasible solutions; descriptive models are used to evaluate decision alternatives or solution proposals respectively and help to get a deeper insight into the problem characteristics. Examples for descriptive models are queueing networks on an analytical and discrete simulation on an empirical basis; constructive models might use mathematical programming techniques or expert knowledge. For production scheduling problems one advantage of descriptive models is the possibility to understand more about the dynamics of the system; with constructive models solutions can be generated. Due to the necessity of short computing times descriptive models seem less applicable to reactive scheduling on a real time level; but finding good answers to reactive problems in advance such should be considered for training purposes of the human scheduler.

There are different possibilities to find solutions for constructive scheduling models [BESW 93]. One is to apply mathematically based "deep" methods (search trees, enumerative or direct methods). Other alternatives are to use the knowledge of domain experts or in a similar sense to build interactive schedulers. Interactive systems should allow the user to guide the search in the directions that he might naturally follow based upon his intuition and knowledge supported by basic functions of the system. All these solution approaches mainly come from the fields of Operations Research (OR) and Artificial Intelligence (AI). In contrast to OR-approaches to scheduling, which are focused on optimization, AI relies on satisfaction, i.e. it is sufficient to generate solutions which are accepted by the problem solver. Disregarding the different solution approaches the status of combinatorial complexity of most of the problems remains the same, as it can be shown that the decision problem is not easier than the corresponding optimization problem [GJ 79]. NP-hard scheduling problems have combinatorially explosive search spaces.

Many efforts of the disciplines of OR and AI are paralleled; the development of dispatching rules, the investigation of bottleneck resources and constraint based scheduling are examples. With priority scheduling as a job-oriented and with bottleneck scheduling as a resource-oriented approach two extremes for rule-based schedule generation exist. There can also be found solution approaches which use exact algorithms mainly for predictive scheduling [BFHS 88]. Especially for reactive scheduling dispatching rules for job release to the system and for job sequencing inside the system are applied [BPH 82, PI 77]. Many AI-based approaches exist which use domain knowledge to solve predictive and reactive scheduling problems [Ata 91].

3. MODELLING REACTIVE SCHEDULING USING FUZZY LOGIC

One main difference between the predictive and the reactive scheduling level is based on the quality and the amount of the input data. For predictive scheduling input data is mainly based on expectations and assumptions on an aggregated level; unforeseen circumstances, like rush orders, machine breakdown, illness of personal staff, etc. can only be considered statistically, if at all. Reactive scheduling uses actual input data. If this is very different from the estimated data, situation-based revisions of previous decisions have to be made.

Reactive scheduling is to be applied if temporary disturbances from the predictive schedule occur. A time dependent strategy in the form of ad-hoc decisions must be devised. If the interruption continues for a longer time, one has to return to predictive scheduling and seek for an alternative strategy on the basis of a new problem instance now with different requirements and possibly different objectives. Again a new strategy has to be found which will then be followed until operations are again disturbed. More details on the integration of reactive and predictive scheduling will be discussed in the next section.

An algorithmic reaction on the reactive level of problem solving based on sophisticated combinatorial considerations is generally not possible because of prohibitive computing times; therefore, the competence of human problem solvers in reaching quality, realtime decisions is extremely important. The human problem solver should be supported by advanced modelling techniques. In order to achieve this we suggest the application of fuzzy logic because it allows to represent the vague, qualitative view of the human scheduler most conveniently. The underlying theory of fuzzy sets [Zad 65] concentrates on modelling vagueness due to common sense interpretation, data aggregation and vague relations. Examples for common sense interpretation in terms of a human production scheduler are long queues of jobs or high machine speed. Data aggregation is used in expressions like skilled worker or difficult situation and vague relations are represented by terms like not much more urgent than or rather equal.

Reactive scheduling requires suitable diagnostic support for quick decision making. This is intended with our approach modelling reactive scheduling by fuzzy logic. The two main components of the model we use are (1) linguistic variables [Zad 73] and (2) fuzzy rules or better decision tables [Kan 86]. A linguistic variable L can be represented by the tuple L=(X,U,f) where set X represents the feasible values of L, set U represents the domain of L, and f is the membership function of L which assigns to each element x $\in$X a fuzzy set $A(x)=\{u,f_X(u)\}$ where $f_X(u)\in[0,1]$. A decision table (DT) consists of a set of conditions (if-part) and a set of actions (then-part). In case of multi conditions or multi actions conditions or actions respectively have to be connected by operators. If all conditions have only one precise value we speak of deterministic DT. In case we use fuzzy variables for representing conditions or actions we also can build nondeterministic DT. These tables are very much alike of how people think. In order to represent the interaction of linguistic variables in DT we have to introduce set-theoretic operations to find the resulting membership function. The most common operations are union ($C=A\cup B$), intersection ($D=A\cap B$) and complement. In case of an union we have $f_C(u)=\max\{f_A(u),f_B(u)\}$, in case of an intersection $f_D(u)=\min\{f_A(u),f_B(u)\}$, and in case of the complement A° of A we have $f_{A°}(u)=1-f_A(u)$. To understand the approach of modelling reactive scheduling by fuzzy logic better consider the following scenario.

There are queues of jobs in front of machines on the shop floor. For each job Jj the number of jobs Nj waiting ahead in the queue, its due date dj and its slack time sj = dj - t are known where t is the current time. Processing times of the jobs are subject to disturbances. Due date and machine assignment of the jobs are determined by predictive scheduling. The objective is to diagnose critical jobs, i.e. jobs which are about to miss their due dates in order to reschedule them. Nj and sj are the linguistic input variables and "becomes critical" is the output variable of the DT. Membership functions for the individual values of the variables are determined by a knowledge acquisition procedure which will not be described here.

The following DT represents the fuzzy rule system.

AND	Small	Medium	Great
Few	soon	later	not to see
Some	now	later	not to see
Many	now	soon	not to see
Very Many	now	soon	later

The rows represent the values of the variable Nj and the columns represent the values of the variable sj. Both variable are connected by an AND-operator in any rule. With the above table 12 rules are represented. Each element of the table gives one value of the output variable "becomes critical" depending on the rule. To find these results the membership functions of the input variables are merged by intersection operations to a new membership function describing the output variable of each rule. The resulting fuzzy sets of all rules are then combined by operations which are applied for the union of sets. This procedure was tested to be most favourable from an empirical point of view.

As a result the decision maker on the shop floor gets the information which jobs have to be rescheduled now, soon, later, or probably not. From this two possibilities arise; either a complete new predictive schedule has to be generated or local ad-hoc decisions can be taken on the reactive scheduling level. Control decisions based on this fuzzy modelling approach and their consequences should be recorded and evaluated, for the purpose of using these past decisions to find better solutions to current problems. Fuzzy expert systems, which should be able to achieve the quality of a self-learning system, could make a significant contribution along these lines.

4. INTEGRATING REACTIVE AND PREDICTIVE SCHEDULING

Production scheduling must be organized in advance; predictive scheduling or offline planning (OFP) is the basis for reactive scheduling or online control (ONC). Due to this production scheduling uses a chronological and functional hierarchy within the decision making steps. This helps to create a presentation of the problem that considers all available information appropriately [Sch 89].

The chronological hierarchy leads - as already mentioned - to the separation of OFP and ONC. Problems are further separated on a conceptual and a specific level in order to produce a functional hierarchy, too. The purpose of the chronological approach to prioritization is to be able to come to a decision through aggregated and detailed modelling, even if future information is inspecific or unavailable. Aside from fulfilling the functional needs of the organization, the basic concept behind the functional hierarchy is to get a better handle on the combinatorial difficulties that emerge from the attempt of simultaneously solving all problems arising in a manufacturing environment. The advantage of such a procedure consists not only in getting a problem-specific approach for investigation of the actual decision problem, but also in the representation of the decision making process within the manufacturing organization [Sch 92].

The rudimentary techniques of solving production scheduling problems presented here work on a closed Analysis-Construction-Evaluation loop (ACE loop). This loop has a feedback mechanism on the levels of OFP and ONC. An overview over the system is shown in the next figure.

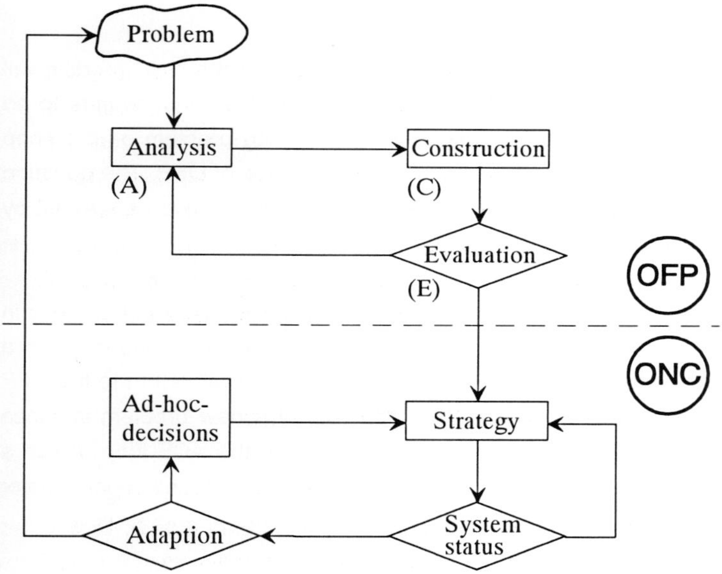

The OFP module consists of an analysis, a construction and an evaluation component. First, the problem instance is analyzed (A) in terms of objectives, constraints and further characteristics. In order to do this the first step for (A) is to describe the manufacturing environment with the scheduling situation as detailed as possible. In a second step from this description a specific model has to be chosen from a set of scheduling models in the library of the system. The analysis component (A) is based upon knowledge-based approaches, such as those used for problems like classification within the realm of Artificial Intelligence.

The problem analysis defines the parameters for the construction (C) phase. From the basic model obtained in (A), a solution for the scheduling problem is generated by (C) using some generic algorithms. The result is a complete schedule that has then to be evaluated by (E). Here the question has to be answered if the solution is implementable in the sense that manufacturing according to the proposed solution meets business objectives and fulfils all constraints coming from the application. If the evaluation is satisfactory to the user, the proposed solution will be implemented. If not, the process will repeat itself until the proposed solution delivers a desirable outcome or no more improvements appear to be possible in reasonable time.

Having generated a feasible and satisfactory predictive schedule the ONC module will be called. This module takes the OFP schedule and translates its requirements to an ONC strategy, which will be followed as long as the scheduling problem on the shop floor remains within the setting investigated in the analysis phase of OFP. The question if the status of the system requires adaptions of the schedule can now be answered by applying fuzzy logic as demonstrated in Section 2.

In case only minor temporary disturbances are diagnosed, a time dependent strategy in the form of an ad-hoc decision can be devised. If the interruption continues for such a long time that a new schedule needs to be generated, the system will return to the OFP module and seek for an alternative strategy on the basis of a new problem instance with new requirements and possibly different objectives within the ACE loop. Again a new ONC strategy has to be found which will then be followed until again major disturbances occur.

As already mentioned, production scheduling problems are changing over time; a main activity of the problem analysis is to characterize the problem setting such that one or more scheduling problems can be modelled and the right method or a combination of methods for constructing a solution can be chosen from a library of scheduling methods or from knowledge sources coming from different disciplines. With this there are three actions to be taken; first the manufacturing situation has to be described, second the underlying problem has to be modelled and third an appropriate solution approach has to be chosen. From this point of view one approach is using expert knowledge to formulate and model the problem and then using "deep"-knowledge from the library to solve it.

The function of OFP is providing flexibility in the development and implementation of desirable production schedules. OFP applies algorithms which can either be selected from the library or may also be developed interactively on the basis of simulation runs using all components of the ACE loop. The main activity of the interaction of the three components is the resolution of conflicts. Whenever the evaluation of some schedule generated by (C) is not satisfactory then there exists at least some conflict between the requirements or business objectives of a problem solution and the schedule generated so far.

The search for a suitable strategy within ONC should not be limited to routine situations, rather it should also consider e.g. breakdowns and their predictable consequences. ONC takes into consideration the scheduling requirements coming from OFP and the current state of the manufacturing system. To that end, it makes the short term adjustments, which are necessary to handle failures in elements of the system, the introduction of new requirements for manufacturing like rush jobs or the cancellation of jobs. An algorithmic reaction on this level of problem solving based on sophisticated combinatorial considerations is generally not possible because of prohibitive computing times of such an approach. Therefore, the competence of human problem solvers in reaching quality, realtime decisions is extremely important. In this sense, the application of fuzzy logic to reactive scheduling seems to be a promising approach.

5. IMPLEMENTATION AND CONCLUSIONS

We have implemented our approach of modelling reactive scheduling by fuzzy logic as a demonstration prototype. Two screens serve as the user interface. On the first screen the jobs waiting in a machine queue and the slack time of the job under investigation are shown. The next picture shows an example problem.

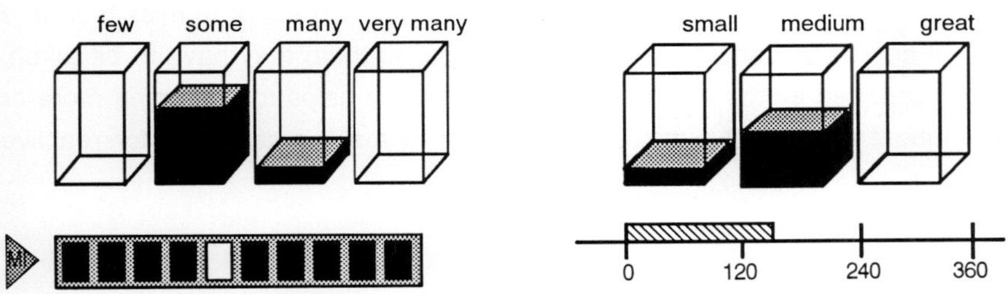

There are ten jobs waiting in a queue to be processed by some machine M_i, the job under consideration J_j is shown by a white rectangle. Above the queue the different fuzzy sets concerning the linguisitc variable N_j and the values of the corresponding membership functions are represented. The slack time s_j of job J_j is currently 130 minutes; again fuzzy sets and membership functions of this linguistic variable are represented above the scale.

Applying the rules of the DT results in a representation which is shown in the next picture. The result of the first part of inference shows that for job J_j the output variable "becomes critical" is related to some positive values for "soon" and for "later" shown by white segments. From this it is concluded by the second part of inference that this job has to be checked again before it is about to be rescheduled.

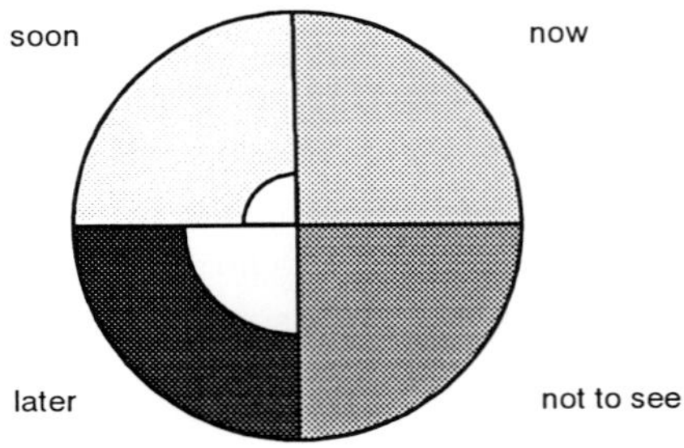

check again

(result of inference)

We demonstrated an approach for the interaction of predictive and reactive scheduling using fuzzy logic. Based on some predictive schedule we consider disturbances on the shop floor resulting in only roughly predictable job processing times. By monitoring all queues of jobs they are analyzed in the sense which jobs are about to miss their due dates. If there are any suitable decision for rescheduling them they have to be taken. This can be supported by the fuzzy logic approach which helps to determine more or less critical jobs and helps the manufacturing staff in making decisions for reactive scheduling.

REFERENCES

[Ata 91]
Atabakhsh, H., A survey for constraint based scheduling systems using an artificial intelligence approach, AI in Eng. 6 (2), 1991, 58-73.

[BPH 82]
Blackstone, J.H., D.T. Phillips, G.L. Hogg, A state-of-the-art survey of dispatching rules for manufacturing job shop operations, Int. J. Prod. Res. 20 (1), 1982, 27-45.

[BESW 93]
Blazewicz, J., K. Ecker, G. Schmidt,, J. Weglarz, Scheduling in Computer and Manufacturing Systems, Springer, 1993.

[BFHS 88]
Blazewicz, J., G. Finke, R. Haupt, G. Schmidt, New trends in machine scheduling, European Journal of Operational Research 37, 1988, 303-317.

[GJ 79]
Garey, M.R., D.S. Johnson, Computers and Intractability: a Guide to the Theory of NP-Completeness, Freeman, San Francisco, 1979.

[Kan 86]
Kandel, A., Fuzzy Mathematical Techniques with Applications, Addison-Wesley, Reading, 1986.

[PI 77]
Panwalkar, S.S., W. Iskander, A survey of scheduling rules, Operations Research 25 (1), 1977, 45-61.

[Sch 89]
Schmidt, G., CAM: Algorithmen und Decision Support für die Fertigungssteuerung, Springer, 1989.

[Sch 92]
Schmidt, G., A decision support system for production scheduling, Journal of Decision Systems 1 (2-3), 1992, 243-260.

[Zad 65]
Zadeh, L.A., Fuzzy sets, Information and Control 8, 1965, 338-353.

[Zad 73]
Zadeh, L.A., The concept of linguistic variables and its application to approximate reasoning, Memorandum ERL-M411, UC Berkeley , 1973.

Knowledge-Based Reactive Scheduling (B-15)
E. Szelke and R.M. Kerr (Editors)
Elsevier Science B.V. (North-Holland)
1994 IFIP

Revisiting the Johnson algorithm for flow-shop scheduling with genetic algorithms

Prof. Hubert Mulkens

Laboratoire de Gestion de la Production, Ecole Polytechnique Fédérale de Lausanne, 1015 - Lausanne, Switzerland

Abstract

The Johnson algorithm is well known for computing an optimized job schedule for a sequence of 2 or 3 machines. However, only heuristic techniques may be applied for more than 3 machines. Among those, new iteratives techniques like genetic algorithms (GA) have proven to be efficient. It is shown here that those techniques may even be better than the Johnson algorithm, for 2 and 3 machines case, due to the numerous equivalent solutions they give to the scheduling problem, hence permitting to take into account other optimisation factors, like the average time in queue. A comparison is also made with the multimachine flow shop CDS and NEH heuristics.

Keyword Codes: G.1.6; I.2.8; I.6.3
Keywords: Numerical analysis, Optimization; Artificial Intelligence, Problem Solving; Simulation and Modeling, Applications

1. JOB SCHEDULING WITH THE JOHNSON ALGORITHM

Johnson did propose an easy to implement scheduling algorithm that minimize the makespan in a flow shop with 2 or 3 machines [1] and did prove that it was not possible to find such an algorithm for more than three machines; for those cases, only heuristics were proposed and among those the CDS and the NEH ones [2,10].

The Johnson algorithm for 2 machines goes as follow (as taken from [3]): select the job with the minimum processing time on either machine 1 or 2; if this time is associated with machine 1, schedule this job first; if it is for machine 2, schedule this job last; remove this job from further consideration; select the job with the next smallest processing time and proceed as for the first one: schedule it second or previous to last and continue this process until all jobs have been scheduled.

When running this algorithm on the example given in Table 1, the computed schedule is 4 2 1 3 and the makespan is 29 units of time (with an average time in queue of 20.5).

Table 1
Application of the Johnson algorithm for 2 machines

Job duration	on machine 1	on machine 2
Job 1	6	4
Job 2	7	7
Job 3	8	3
Job 4	5	6
Computed schedule (Johnson algorithm): 4 2 1 3 with a makespan of 29		

For 3 machines, the algorithm is very similar: it is equivalent to a two machine problem where the first machine is a compound of machines 1 and 2 and the second machine is a compound of machines 2 and 3. However, this algorithm does only work if no job duration on the second machine exceed any job duration on the first or the last machine.

When running this algorithm on the example given in Table 2., the computed schedule was found to be 3 1 2 4 with a makespan of 32 and an average time in queue of 21.8 units of time.

Table 2
Application of the Johnson algorithm for 3 machines

Job duration	on machine 1	on machine 2	on machine 3
Job1	5	3	6
Job2	7	2	9
Job3	4	2	4
Job4	6	3	5
Computed optimal schedule (Johnson algorithm): 3 1 2 4 with a makespan of 32			

2. WHAT ARE GENETIC ALGORITHMS (GA)?

Genetic Algorithms were invented in 1975, by J. Holland at the University of Michigan [4], by mimicking nature.

A combinatorial optimisation problem consist in finding the value for several parameters that minimize (or maximize) a predefined cost function.

In order to do that with GA, we have first to code the serie of parameters to optimize: this will be the chromosome. Hence, the chromosome will be a serie of bits made of the juxtaposition of varying length bit strings, each string coding one parameter. The order and the length are fixed for all parameters: the same parameter will be found at the same place on all chromosomes.

Then, starting from a population of chromosomes, where each of them represents the same serie of parameters but with different values, some evolution rules will be applied in order to arrive at a population that is more "fit", i.e. where the parameters will have reached their optimal value.

It is then necessary to define a cost function for each chromosome. This cost function will somehow depend on the values of the parameters.

The evolution is performed basically by two operators: the crossover operator and the mutation operator.

The crossover operator will cut two chromosomes in half and recombine the corresponding halves by crossing them between the two original chromosomes. The cut can only be made between two strings coding for two different parameters, not in the middle of a parameter.

The mutation operator will randomly hit one bit in the chromosome and flip it. The rate of mutation has to be adjusted in order for the mutation to enrich the population with new mutant chromosome but without perturbing the convergence process towards an optimum (or near optimum) solution.

The dynamic of the algorithm is created by the following iteration:

1-first generate a random population of chromosome

2-evaluate each chromosome

3-apply the crossover operator to any pair of chromosome of that population: this double temporarily the population size,

4-evaluate the children created

5-select a new population (allow some members to die: either the parent population or the least efficient chromosome) in order to keep the population size constant,

6-apply the mutation operator on some randomly selected chromosomes,

7-restart at point 2 if a stopping criteria is not yet met. This stopping criteria may be a number of iteration or the value of the fitness function on one or all the chromosomes.

Genetic algorithm have proven to be very efficient in parameter optimisation, classifier problem and learning problem.

We will apply them to scheduling.

3. JOB SCHEDULING WITH GA

The application of GA to scheduling problems does not follow all the rules of classical GA [6].

The chromosome will code the order we want to optimize. Hence a simple representation such as a vector of integer, is a sufficient and efficient representation.

The crossover operator has to be modified in order to avoid the mixup of the order vectors: classical crossover will create vector with double order number and missing ones. We did apply the crossover operator found in [7-p.79] or the PMX operator [5].

The mutation operator, for the same reason cannot simply change one order number into another one. Hence we did use the partial sublist permutation operator [7-p.81].

The fitness function will be related to the ordering problem we try to optimize. This will often be a minimization problem: for example, in manufacturing, we want to minimize the total time needed to make a serie of jobs. This total elapsed time between the start of the first job to the finish of the last one of the serie, is called the makespan. The fitness function will often be the computed makespan or measure related to it.

The fitness function computation will also contain all the constraints of the problem. Violation of some constraints may be taken into account by giving heavy weigth to those in the fitness (hence not favorable in a minimization problem).

We made some specific tests to try to optimise the use of GA for scheduling problems. To be more specific, we did apply GA on an ordering problem with a well known and computable solution [7]. By comparing the solution with the results obtained with GA, it was possible for us to set the parameters of our genetic algorithm:

-the population size n was choosen as twice the length of the vector to order (in other words, twice the number of jobs to schedule)

-the initial population is made of n randomly defined order vectors

-the crossover operator was applied between pairs of randomly selected individuals (order vectors) according to their fitness

-the population used at every new generation was made of the n best individuals from the parent and child population (hence, we do not keep all child)

-the mutation operator was used with a rate of 10%: higher rate were found to destroy the convergence mechanism and lower rate imply the risk of missing good solutions.

-the stopping criteria will be when we return for the tenth times to a population whose global fitness (i.e. sum of fitness value of each chromosome) is the minimum found up to now.

4. APPLICATION OF GA TO THE 2 MACHINES SCHEDULING PROBLEM

We did apply our genetic algorithm to the 2 machines problem given in Table 1. The results are reported in Table 3.

Table 3
Application of GA to the 2 machines scheduling

Initial random population of 8 scheduling vectors
Convergence to a stable minimum of the makespan criterion in 17 generations
Value of the makespan: 29 units
Final population:
2 vectors with the schedule: 4 2 1 3 (average time in queue of 20.5)
6 vectors with the schedule: 2 1 4 3 (average time in queue of 21.2)

We observe that the algorithm did rapidly converge to the correct solution (the makespan is the same as the one given by the Johnson algorithm) and that the found schedules do comprise the one given by the Johnson algorithm and also another one showing the same mimimum makespan and a different average time in queue.

We did try several examples: for every one of them, the makespan found by our genetic algorithm was the same as the one found by the Johnson algorithm but we did not always found a schedule in the final population that was the same as the one given by Johnson's algorithm. However, for several examples, we found other schedules than the Johnson's one with a smaller value of the jobs average time in queue (see for example Table 4).

Table 4

Comparison of Johnson and GA on a 10 job schedule

Job duration	on machine 1	on machine 2
Job 1	5	3
Job 2	7	2
Job 3	4	5
Job 4	6	6
Job 5	1	9
Job 6	9	1
Job 7	3	3
Job 8	2	3
Job 9	7	8
Job 10	10	2

Johnson's algorithm: makespan: 55 and average time in queue: 30.2

Johnson's schedule: 5 8 7 3 4 9 1 10 2 6

Genetic algorithm: Initial random population of 20 scheduling vectors

Convergence to a minimum of the makespan in 27 generations - makespan: 55

Final population: (all made of schedules of makespan of 55 units)

Average time in queue -	schedule
35.5	10 5 8 3 9 2 1 7 4 6
36.2	9 3 4 5 10 2 8 1 7 6
32.8	5 9 4 3 7 8 10 1 2 6
39.0	2 9 5 3 4 8 10 1 7 6
33.9	3 4 5 7 9 10 8 1 2 6
33.9	8 9 4 3 5 1 2 10 7 6
33.8	8 9 5 1 4 10 3 2 7 6
37.0	9 4 5 3 2 7 1 8 10 6
36.4	4 9 5 2 3 7 1 8 10 6
35.8	9 1 5 4 10 3 2 8 7 6
30.4	5 3 2 1 7 9 8 4 10 6
33.8	8 9 5 1 4 10 3 2 7 6
31.5	5 7 9 3 4 2 1 8 10 6
36.9	9 4 5 1 3 10 8 7 2 6
38.9	2 9 5 3 4 10 8 1 7 6
37.1	9 5 7 4 3 8 1 10 2 6
35.3	7 9 5 3 4 10 1 8 2 6
31.1	5 9 7 3 2 8 1 4 10 6
37.2	9 5 4 1 7 3 8 10 2 6
30.0	5 7 2 3 1 9 8 4 10 6

The Johnson schedule is not in the final population

The last schedule of the population has a shorter average time in queue: 30.0 (in place of 30.2)

We made other trials with up to 20 jobs to schedule: we found no convergence problem (neither no convergence i.e. instable solution nor premature convergence to a wrong solution i.e. one with a makespan which is not the smallest possible).

5. APPLICATION OF GA TO THE 3 MACHINES SCHEDULING PROBLEM

The application of our GA scheduling to the problem given in Table 2, as solved with the Johnson algorithm, did lead to the result given in Table 5.

Table 5
GA application to the 3 machines scheduling problem

Initial random population of 12 scheduling vectors
Convergence to a stable minimum of the makespan criterion in 15 generations
Value of the makespan: 32 units
Final population:
7 vectors with schedule 1 3 2 4 (and average time in queue of 22.8)
4 vectors with schedule 3 1 2 4 (and average time in queue of 21.8)
1 vector with schedule 1 2 4 3 (and average time in queue of 24.2)

We observe that with GA we do obtain the same makespan as with the Johnson algorithm, but the algorithm give us 3 equivalent schedules (including Johnson's one).

We did try several other examples: for every one of them, the makespan found by our genetic algorithm was the same as the one found by the Johnson algorithm but we did not always found a schedule in the final population that was the same as the one given by Johnson's algorithm. However, for several examples, we found other schedules than the Johnson's one with a smaller value of the jobs average time in queue (see for example Table 6).

Table 6
Comparison of Johnson and GA on a 10 job schedule on 3 machines

Job duration	on machine 1	on machine 2	on machine 3
Job 1	5	3	10
Job 2	7	2	6
Job 3	5	4	5
Job 4	6	4	6
Job 5	8	1	9
Job 6	9	1	5
Job 7	5	3	9
Job 8	9	3	7
Job 9	7	2	8
Job 10	10	2	6

Johnson's algorithm:
Johnson's makespan: 79
Johnson's schedule: 1 7 3 5 9 4 8 10 2 6
Johnson's average time in queue: 50.5

Table 6 (continuing)
Comparison of Johnson and GA on a 10 job schedule on 3 machines

Genetic algorithm:

Initial random population of 30 scheduling vectors
Convergence to a stable minimum of the makespan criterion in 28 generations
Value of the makespan: 79 units
Final population: (all made of schedules of makespan of 79 units)

Average time in queue -	schedules
49.3	7 9 5 3 2 4 1 8 10 6
50.8	7 5 1 9 4 2 3 6 8 10
50.5	1 7 2 5 9 4 3 6 8 10
50.4	7 5 1 4 2 3 9 8 10 6
49.0	7 2 1 4 9 3 5 6 8 10
49.3	7 2 9 4 1 8 5 3 10 6
49.8	7 5 1 4 6 8 3 2 9 10
49.3	7 2 4 1 8 5 9 3 10 6
49.4	7 2 1 4 9 5 3 6 8 10
50.4	7 5 9 1 3 4 2 6 8 10
50.6	7 5 1 4 9 2 3 6 8 10
48.8	7 5 9 2 4 6 3 1 8 10
50.4	7 4 1 9 5 8 2 3 6 10
48.5	7 3 5 4 9 2 1 8 10 6
48.5	7 3 5 4 9 2 1 8 10 6
48.5	7 3 5 9 4 8 2 1 10 6
49.4	7 2 1 4 9 5 3 6 8 10
50.6	1 5 2 7 9 3 4 10 8 6
49.5	7 9 5 4 2 3 1 8 10 6
50.0	1 2 5 7 3 9 4 10 8 6
49.9	1 7 2 4 9 5 3 6 8 10
50.6	7 5 1 4 9 2 3 6 8 10
49.7	7 2 1 4 8 5 9 3 10 6
50.6	7 5 1 4 9 2 3 6 8 10
49.9	1 7 2 4 9 5 3 6 8 10
49.2	7 4 1 2 9 8 3 5 6 10
50.5	7 1 4 5 9 3 2 10 8 6
50.0	7 5 10 1 3 9 4 2 8 6
50.4	7 9 5 4 1 10 2 3 8 6
49.9	7 5 10 1 2 3 9 4 8 6

The Johnson schedule is not in the final population
23 found schedules have a shorter average time in queue than with the Johnson algorithm.

(Note: in the GA examples with 3 machines, we did increase the population size from twice the length of the vector to order to the trebble of that length; this has no impact on the result, it is simply an exercise to show the increase in solution richness).

Finally we tried to relax the constraint put by Johnson in order to succeed in finding an optimal schedule (i.e. having the second machine always dominated by the first or the third).

This was also tested by Wagner and Giglio [9]: they kept using the Johnson algorithm and they did found that an optimal solution was found in half the cases they tried. We did investigate this question with our GA approach (see Table 7).

Table 7
Comparison of "relaxed" Johnson and GA for 3 machines

Job duration	on machine 1	on machine 2	on machine 3
Job 1	5	7	10
Job 2	2	7	8
Job 3	10	2	8
Job 4	2	11	6
Job 5	4	6	7
Job 6	8	10	7

Computed scheduling: 2 5 1 4 6 3
Makespan: 58 (with average time in queue of 37.2)
Application of GA to an initial random population of 18 vectors
Convergence in 26 iterations
Best schedule: 2 1 4 5 3 6 (all vectors do converge on this one)
Makespan: 55 (with average time in queue of 36.7)

In Table 7 we show one example where GA gave a better result than Johnson: the difference is 5.2%. But contrary to the observation of Wagner and Giglio, we did observe this superiority of GA over Johnson in only 20% of our generated test cases (Wagner and Giglio found approximatively 50%) and in all other cases, the results were the same for the makespan, and better for GA when considering the average time in queue.

6. COMPARISON OF GA WITH THE CDS MULTIMACHINE HEURISTIC.

The Johnson algorithm is known to be limited to the flow shop scheduling of 2 or 3 machines. Several heuristics were developped in order to propose near optimal solutions for n machines, with n > 3. Among them two are considered to be the most efficient: the CDS algorithm [2] and the even better NEH algorithm [10]. We did investigate the behaviour of our GA method compared to CDS and NEH. The results are shown in Table 8 and Table 9.

Table 8 presents some specific results. We did not conduct an exhaustive test (only 10 samples for each problem and we give some characteristics results here) but we have observed that for the first problem (5 jobs-10 machines) our GA method always gave the best results which was also given either by CDS or NEH. For the second problem (8 jobs-10 machines) we have observed that in one experiment NEH gave a better result (with less than 1% of difference) and for all others GA was the best (from 0.8% to 1.5%). In that second problem, we also do observe than the number of equivalent good schedules is bigger than one, hence enabling the choice of the best schedule according to some other criteria.

Table 8
Comparison of GA, CDS and NEH for the n jobs m machines problem.

a) Application to a 5 jobs-10 machines problem

GA				CDS		NEH		best result
Number of iterations	Number of good schedules	make span	average time in queue	make span	average time in queue	make span	average time in queue	
40	2	108	86.2	109	86.2	108	86.2	GA,NEH
42	1	105	84.6	105	84.6	107	84.2	GA,CDS
40	1	103	86.2	103	86.2	103	86.2	All
40	1	104	88.6	105	88.0	104	86.2	GA,NEH
40	1	98	82.4	101	82.6	98	82.4	GA,NEH

b) Application to a 8 jobs-10 machines problem

GA				CDS		NEH		best result
Number of iterations	Number of good schedules	make span	average time in queue	make span	average time in queue	make span	average time in queue	
61	3	129	100.1	130	102.5	128	98.6	NEH
52	8	128	99.2	130	100.8	130	100.9	GA
64	1	132	101.0	135	103.4	132	101.0	GA,NEH
47	4	130	99.0	133	99.6	132	109.8	GA
54	1	125	94.8	126	94.1	128	99.5	GA

In Table 9 we give the starting data for the example of the second experiment of the 8 jobs-10 machines problem.

Table 9
Details on one comparative run between GA, CDS and NEH

Machines:	1	2	3	4	5	6	7	8	9	10	Total
Job1	8	8	2	1	9	11	1	2	12	2	56
Job2	11	3	7	10	6	2	10	7	10	11	77
Job3	6	9	8	6	6	4	11	7	8	10	75
Job4	1	10	11	8	11	8	9	11	6	2	77
Job5	3	9	5	7	10	2	7	8	10	2	63
Job6	8	2	11	6	4	6	7	8	10	7	69
Job7	6	6	6	10	3	5	2	10	9	4	61
Job8	10	2	7	6	2	8	7	2	6	3	53

Use of the CDS algorithm

Computed schedule 6 5 7 2 3 1 8 4

Makespan : 130 (with average time in queue of 100.8)

Use of the NEH algorithm

Computed schedule 5 6 1 7 3 2 8 4

Makespan: 130 (with average time in queue of 100.9)

Table 9 (continuing)
Details on one comparative run between GA, CDS and NEH

Use of GA

Size of the population: 24 schedules

List of best equivalent schedules (8 solutions, all better than CDS or NEH)

5 7 3 6 1 2 4 8	makespan: 128	(with average time in queue 99.2)
5 6 3 1 7 2 4 8	makespan: 128	(with average time in queue 99.5)
5 7 6 3 1 2 4 8	makespan: 128	(with average time in queue 99.5)
5 7 3 1 6 2 4 8	makespan: 128	(with average time in queue 99.5)
5 6 7 3 1 2 4 8	makespan: 128	(with average time in queue 99.5)
5 6 3 7 1 2 4 8	makespan: 128	(with average time in queue 99.6)
5 7 3 2 1 6 4 8	makespan: 128	(with average time in queue 100.0)
7 5 6 3 1 2 4 8	makespan: 128	(with average time in queue 100.5)

7. INFLUENCE ON THE NUMBER OF JOBS AND MACHINES ON THE GA CONVERGENCE

It has been conjectured [10] that a problem with more machines than jobs was easier to solve than the contrary. We do confirm this conjecture when using GA. We made a numerical experiment where we generated several runs having to deal for one half with the scheduling of 5 jobs on 10 machines and on the other half with the scheduling of 10 jobs on 5 machines. The results are reported in Table 10.

It takes nearly three times as much in the number of iterations and a rapid increase in the richness of the obtained solutions is also noteworthy (a factor 10), when dealing with more jobs and less machines.

Table 10
Investigation of jobs/machine scheduling difficulty

Number of jobs	Number of machines	Average number of iterations	Average number of different good schedules	Average number of efficiency
5	10	38.4	1.2	89.2 %
10	5	110.4	10.4	85.7%

8. IMPLICATIONS OF GA FOR REACTIVE SCHEDULING

More and more in discrete manufacturing, the workshops tend to be organized as flow shops: material move, job dispatching, shop floor control is all made easier. The success of cellular manufacturing is there to confirm this necessity to transform a job shop into a flow shop. At the same time however, the work scheduling is still done in practice with the help of very crude rules (critical ratio e.g.).

Nowadays competitive advantages will come with a change of scope: scheduling should move from a data driven approach (made of static lists of jobs to schedule) to an event driven approach (where the list of jobs to schedule becomes dynamic: jobs are cancelled, other become more urgent, new jobs have to be taken into account...). In this respect, the workshop scheduling made with the help of GA is very powerful for two important reasons:

- it is straightforward to recompute a new schedule by taking into account the modifications in the job list,
- the fitness function used to determine the quality of a schedule may be easily modified to take into account new (temporary) constraints applying to the jobs to schedule.

9. COMPUTER SUPPORT

All the computations were made by using a specific software written in Symantec Pascal on an Apple Macintosh Powerbook 170. The software was easy to develop, code and debug. It consists of less than 2'000 lines of code. A typical run lasted from a few seconds (optimisation of 4 jobs on 2 machines) to less than five minutes for the optimisation of 20 jobs on 3 machines. It should be understood that this hardware configuration is far from ideal for this kind of software, which need much more CPU power.

10. CONCLUSIONS

The application of genetic algorithms to the machine scheduling problem was proven to be very efficient:

-in all sample case tested, for the 2 and 3 machines problem, our GA program came out with a population of schedule vectors with the same makespan as the one obtained with the direct application of the Johnson algorithm,

-within this population, we found several other schedules than the single one given by the application of the Johnson algorithm,

-among those equivalent schedules w.r.t. the makespan, it is possible to single out one or more schedules that are better than the Johnson one when considering other parameters like the average job time spent in the workshop (the mean tardiness could also be tested if due dates were given),

-more generally, this gives more flexibility to the real scheduling of jobs in a workshop, because the operator may take into account other constraints (like for example tooling change, of material handling); we do consider this as an important point in order to make this kind of software a true scheduling tool accepted and used on the shop floor,

-the comparison with the CDS and the NEH heuristics drive us to the same conclusion: GA is as efficient but more rich in the number of possible equivalent solutions,

-last but not least, this kind of algorithm is easy to implement and fast to run, hence it has a real value for flow shop schedulers.

Those results should be considered as preliminary results of the use and integration of GA into scheduling software for real workshops. We are currently working on an implementation of this software as a stand alone tool in a workshop.

REFERENCES

1. S.M. Johnson - "Optimal two and three stages production schedules with setup time included" - Naval Research Logistics Quarterly 1 (1954) p.61
2. H.G. Campbell, R.A. Dudek, M.L. Smith - "A heuristic algorithm for the n job m machine sequencing problem" - Management science 16,no 10, (1970)
3. T. Vollmann, W. Berry, D. Clay Whybark - "Manufacturing Planning and Control Systems" - 3d éd.Irwin - 1991 - p.536
4. John Holland - "Adaptation in natural and artificial systems" - The University of Michigan Press - Ann Arbor - 1975
5. D. Goldberg & R. Lingle - "Alleles,loci and the traveling salesman problem" - Proc. Int. Conf. on Genetic Algorithms and their applications - 1985 - p154
6. G.E. Liepins & M.R. Hilliard - "Genetic Algorithms: foundations and applications" - Annals of Operations Research, 21 (1989) p.31
7. Lawrence Davis ed. - "Handbook of genetic algorithms" - Van Nostrand Reinhold - 1991
8. H. Mulkens - "Une note sur le réglage des paramètres dans l'application de l'algorithmique génétique aux problèmes de séquencement" - internal paper EPFL-LGP 92/8
9. R.J. Giglio & H.M. Wagner - "Approximate solutions to the 3 machine scheduling problem" - Operations Research 12,nr 2 (1964)
10. M. Nawaz, E.E. Enscore, I. Ham - "A heuristic algorithm for the m-machine n-job flow shop sequencing problem" - Omega 11 (1),p.91 (1988)

Knowledge-Based Reactive Scheduling (B-15)
E. Szelke and R.M. Kerr (Editors)
Elsevier Science B.V. (North-Holland)
1994 IFIP

Scheduling for Absolute Temporal Consistency

P. Raja, L. Ruiz, J. Hernandez, G. Noubir, M. Riese

Laboratoire d'Informatique technique
Swiss Federal Institute of Technology
IN-Ecublens, CH 1015 Lausanne, Switzerland
email: raja@litsun.epfl.ch

ABSTRACT

We address temporal consistency issues in a real-time system. The terms absolute and relative consistency characterize the temporal consistency. A function called *age* is introduced to express the temporal knowledge. It contains knowledge about the rate at which a process variable evolves. A scheduling problem with absolute temporal consistency as a scheduling constraint is formulated. For the dynamic scheduling in the occurrence of unpredictable events, the application of reactive scheduling techniques is briefly discussed. Problems that arise when such techniques are used in the presence of real-time constraints are highlighted.

Keyword Codes: I.2.4; I.2.9; I.2.m

Keywords: Knowledge Representation Formalisms and Methods; Robotics; Miscellaneous

1. INTRODUCTION

Control applications run at the lowest level of factory automation, namely on the floor shop. Such applications are in constant interaction with an external physical process through field devices (e.g. sensors, actuators, local controllers). Their major activity is to read values (from sensors), calculate new commands (by controllers) and send new commands (to actuators). Generally, these three steps are carried out either in a *cyclic and periodic* manner [1, 3, 4], or in a *reactive way*, meaning that the application must react to sudden changes (events) in the environment [2]. The former is often referred to as state-based or sampled systems and the latter as discrete event systems. In both cases, a time limit is specified within which the application must take control decisions.

In computer control systems theory [1], the external process is generally modeled as a set of process variables that evolve as time progresses (i.e. they are functions with time as one of their parameters). Depending on the type of application, a process variable can be temperature, pressure, position, etc. The state of a variable at an instant t is given by its value at t. A variable can be either continuous (for example temperature) or discrete (for example open/close valve). On the other hand, within the control system it is always represented in its discrete form. Therefore, we have two different representations of a process variable; an *external representation* within the external process and a corresponding *internal representation* within the control system. Not all values taken by a process variable can be used internally by the control system for several reasons such as: the differences in their physical characteristics (continuous and discrete); the complexity that would otherwise arise if all external values were to be used by the control system. Still both representations are semantically and temporally related with each other. Their temporal relationship is what is called as *data temporal consistency* or simply *temporal consistency.* Two types of temporal consistency must be distinguished: the first, when a value is used to determine the state of a variable; and the second, when a set of values is used to determine the (entire or partial) state of the external process.

1.1. Related Work and Main Contributions

Representing time and time related constraints has been an active area of research within Artificial Intelligence [11,12,13,14,15]. In [13], the author classifies temporal knowledge as absolute and relative. The term absolute temporal knowledge is used to represent knowledge that explicitly makes references to periods of time whereas relative knowledge does not make any such explicit reference. Even in real-time systems, a key issue consists of modeling time and time related application semantics. Temporal consistency expresses an important characteristic of real-time applications that control a rapidly evolving external environment.

The initial papers addressing data temporal consistency issues in real-time systems are [7, 8, 9, 10]. In previous papers, we have concentrated on characterizing the temporal consistency [3, 4, 5, 6]. We consider temporal consistency as defined in [9]. Two types of temporal consistency relationships are distinguished, namely the *absolute* and *relative consistency.* In order to explicitly incorporate and enrich the temporal knowledge required for absolute consistency, a function called *age* is introduced. The age function characterizes the rate at which a value ages. This paper discusses the different aspects and implications of the age function.

After defining temporal consistency and discussing its significance for state-based and discrete event systems, we derive scheduling constraints for the absolute temporal consistency condition based on the age function.

The scheduling problem is formulated for both types of control systems. For periodic tasks (i.e. state-based systems), *a priori* schedules can be constructed. On the other hand, for discrete event systems, constructing *a priori* schedules does not yield in efficient schedules as the arrival times of the events are not known. The former can be seen as a static scheduling and the latter as a dynamic scheduling problem. In several scheduling problems for manufacturing systems, similar classifications between static and dynamic scheduling are made. The difference between the solutions employed for these systems and the real-time systems lies in the response times requirements of the application. As far as dynamic scheduling is concerned, due to the short response times, only *reactive scheduling* techniques are well-suited. This means that scheduling decisions are taken only when events occur. Finally, practical aspects of the scheduling problem are discussed.

In Artificial Intelligence, the scheduling problem has also been addressed in conjunction with the job-shop scheduling problem [11, 15] and also real-time scheduling [14]. Temporal requirements are generally formulated as scheduling constraints. Because of the complexity of such constrained scheduling problems (generally NP-complete), heuristic solutions are often adapted. In [11], the different ways in which such scheduling problems can be formulated and possible solutions are proposed. Knowledge-based scheduling techniques are used in [15]. For real-time systems, several of these approaches are interesting. However, the performance of these techniques have to be evaluated in a different way for real-time systems. A few possibilities that can be used as a base for defining performance measures are discussed.

1.2. Organization

In Section 2, the concept of temporal consistency is explained and the age function is introduced. In Section 3, a graph model representing the control application is presented and the various aspects of the scheduling problem are discussed. Section 4 concludes the paper.

2. TEMPORAL RELATIONSHIPS

Let us examine the two types of temporal relationships identified in Section 1 more closely. The first one defines a relationship between the external and internal representation of a process variable and is called *absolute temporal consistency.* The second one defines a relationship between two or more internal representations within the control system and is called *relative temporal consistency* (shown in Fig.1). The terms employed to describe the two types of relationships are *accuracy* and *precision* in [8] and *age* and *dispersion* in [9]. These terms will now be explained with an example.

2.1. Age-function

We now introduce *age* function that defines the rate at which a process variable evolves in time. It tells us how fast a value *ages* and therefore, how fast a value can be consumed. The age function respects the following properties:

$$age(a,t) = \begin{cases} 0 & t = 0 \\ f_a(t) & otherwise \end{cases}$$

where $f_a(t)$ is a monotonous function, i.e. $t_1 < t_2 \Rightarrow f_a(t_1) \leq f_a(t_2)$. This means that a value can only become *older* as time progresses.

2.2. Absolute and Relative Temporal Consistency

Consider two process variables a, b and let $[a,t_a,v_a]$ and $[b,t_b,v_b]$ be their internal representations, where t_a, t_b indicate the instant when the values v_a, v_b of a, b are acquired (e.g. read by sensors). At instant t, a and b are said to be *absolutely consistent* if

$(age(a,t)-t_a \leq A_a)$ and $(age(b,t)-t_b \leq A_b)$ for a given A_a, A_b (Absolute consistency condition).

Similarly, a and b are said to be *relatively consistent* at t if

$|t_a-t_b| \leq R$ for a given R (Relative consistency condition).

A_a, A_b is called the *absolute* and R the *relative threshold.*

Therefore, the absolute temporal consistency tells us whether a value is too *old* or not (whether it can still be used to take any valid control decision or not), and the relative temporal consistency tells us whether a set of values can be used to form a *snapshot image* of the external process. Note that the age of a value does not directly play any role in determining the relative consistency.

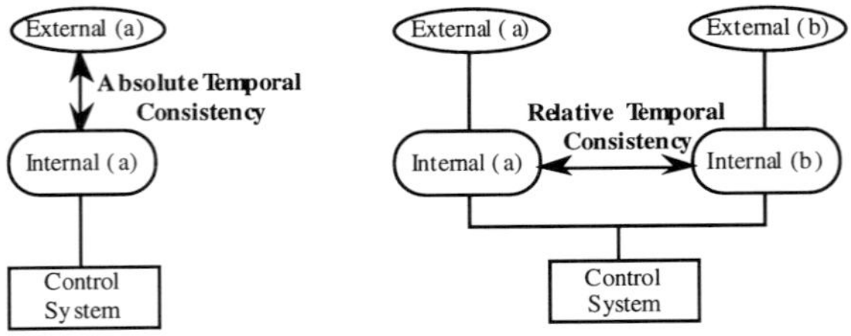

Fig. 1. Absolute and relative temporal consistency of a process variable

An Illustrative Example

Consider two sensors that measure position in x- and y-direction of a moving robot arm. If the measurements are not taken at the same time, then it is not possible to determine the exact position of the arm based on the measured values. Therefore, these two values must be relatively consistent. Now as time progresses and the arm has moved, the two values cannot be used by the control system anymore as they do not correspond to the current position of the arm. Consequently, the values though relatively consistent are not absolutely consistent.

2.3. Implications for state-based and discrete event systems

As we have seen, temporal consistency contains state and temporally related information. State-based systems rely on the relative consistency more than the absolute consistency because (by definition) their operations are based on the state of the external process. However, absolute consistency is also required for the control system to decide whether valid decisions can still be taken in real-time. Absolute temporal consistency, on the other hand, is more relevant to discrete event systems because the control system should know when an event occurred.

Another important characteristic of temporal consistency is with respect to the ordering property. If we know that a set of variables are relatively consistent, then we cannot determine the order in which the values were acquired. The ages of the values are needed for this purpose. The order in which the events occur can be relevant in discrete event systems but not in state-based systems.

Finally, the age is a rich function and can be used in several applications. For example, if the age of a value is greater than a specified threshold, then exceptional conditions can be defined. In the next section, this will be shown for a periodic task.

In what follows, we concentrate on the absolute consistency. The scheduling problem is formulated with the absolute consistency as a scheduling constraint.

3. SYSTEM MODEL

The entire control system consists of a set of tasks and is represented as a graph. The system is modeled as a finite directed graph $G=(P, E)$, where the set of nodes P corresponds to tasks, and the set of edges E represents data dependency between the tasks. For our control system, the field devices and controllers are the tasks.

We assume that there are n tasks in the system and therefore, the task set is $P=\{p_1,p_2,...,p_n\}$. The nodes are assumed to be arbitrarily connected, where the edge set connecting the tasks is given as $E=\{e_1,e_2,...,e_m\}$. We shall write $p_i \xrightarrow{e_k} p_j$ to denote an edge e_k connects task p_i with p_j. e_k is the *outgoing edge* of p_i and the *incoming edge* of p_j.

3.1. Subsets of the Graph

Temporal consistency is expressed for k subsets of nodes, $P_1, P_2,..., P_k$. A task subset is an ordered tuple of tasks denoted as $P_i=(p_0^i, p_1^i, ..., p_{r_i}^i)$. We shall call p_0 as the consumer and $p_1,p_2,...,p_{r_i}$ as the producer tasks. For example, the producer tasks are the sensors and the consumer tasks are the controllers and the actuators.

An Example

This example illustrates how the task subsets are formed. Fig.2 shows an example with $P=\{p_1,p_2,...,p_8\}$ and $E=\{e_1,e_2,...,e_9\}$. In this example, the node subsets are $P'=\{(p_4,p_1,p_2),(p_5,p_2,p_3),(p_6,p_4),(p_7,p_4,p_5),(p_8,p_3,p_5)\}$, $P_1=(p_4,p_1,p_2)$, $p_0^1=p_4$, $p_1^1=p_1$, $p_2^1=p_2$, and $r_1=2$.

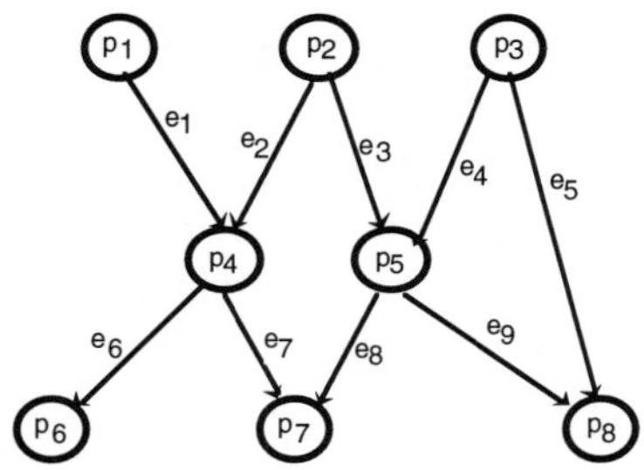

Fig.2. An example with 8 tasks

3.2. Static Scheduling Problem (for periodic tasks)

Now we are able to formulating the scheduling problem for a set of periodic tasks executing on a uniprocessor.

The behavior of a periodic task, $p=(\varepsilon,c,T)$ is defined as follows.

i) ε is the first instant when the task executes for the first time (ε is called as the relative phasing of the task).

ii) The $(k+1)^{th}$ invocation of the task will occur at time $t_{k+1} = t_k+T$. (T defines the period time of the task).

iii) The kth execution of the task must begin no earlier than t_k and must be completed no later than t_k+T.

iv) c is the computation time of the task, i.e. if the task starts its execution at instant t_k then it will complete its execution at instant t_k+c. c must be less than the period time T.

Generally, it is assumed that the computation times for the sensors are negligible [9] (i.e. $c=0$). Consequently, for a given subset (with producer and consumer tasks) only the computation times of the producer tasks are equal to set.

Given

- The control system modeled as a graph of tasks containing subsets $P_1, P_2, \ldots, P_k$, such that $P_i = (p_0^i, p_1^i, \ldots, p_k^i)$.

- The periodic behavior of the producer tasks $1 \leq i \leq k$: $p_i=(\varepsilon_i, 0, T)$, (all producers have the same period time).

- The consumer task specified as $p_i=(\varepsilon_0, c_0, T)$, (only the computation time of the consumer task is given).

- The age function of every process variable a_i produced by the producer task p_i as follows:

$$age(a_i,t) = \begin{cases} 0 & \text{for } t = \varepsilon_i \\ f_{a_i}(t-((k-1)T+\varepsilon_i)) & \text{for } kT \leq t < kT+\varepsilon_i \quad (k \geq 1) \\ f_{a_i}(t-(kT+\varepsilon_i)) & \text{for } kT+\varepsilon_i < t < (k+1)T \quad (k \geq 1) \end{cases}$$

The above age function says that the age of a_i is 0 at instants when a new value is acquired by the sensor $(t=kT+\varepsilon_i)$ and the value *ages* monotonously otherwise. It is a discontinuous function with a jump at instant $kT+\varepsilon_i$.

The function $age(a_i, t)$ is periodic. The maximum value of the age function for a variable a_i is $f_{a_i}(T)$.

Now another application of the age function (that was mentioned in the last section) can be seen. Let a_i be the value acquired periodically by a sensor task. If the age of a value is greater than $f_{a_i}(T)$, then it means that the sensor was not able to acquire the value, for example, because of a fault. One can define that if the age of a value is *older* than a given threshold, then a sensor failure has occurred. Thus, the age function can be used for the detection of sensor faults.

Required

The instants at which the consumer tasks $p_0^1 \in P_1, p_0^2 \in P_2, \ldots, p_0^k \in P_k$ can be scheduled on a uniprocessor (in a non-preemptive way), such that the absolute consistency condition is respected by all of them. In other words, we need to know the relative phasing, ε_0^i of consumer task p_0^i, for $1 \leq i \leq k$.

In what follows, we shall consider only one subset and therefore, simplify the notation. The ith subset will be denoted as $P_i = \{p_0, p_1, \ldots, p_k\}$.

The time interval within which a consumer task p_0 can consume a value is the intersection of the time intervals within which the values required by the consumer task are valid. This is illustrated in Fig.3. The consumer task p_3 consumes two values from producer tasks p_1 and p_2. The figure shows the age functions of the two values and the absolute thresholds are also depicted. Observe that the value from producer p_1 can be consumed between the interval $[T+\varepsilon_1, r_1]$ and the value from p_2 can be consumed between the interval $[T+\varepsilon_2, r_2]$. However, as p_1 has to consume both the values, the scheduling zone of p_3 is only the intersection of the two intervals.

The following observations can be made:

- because the producer tasks are all periodic, the intervals within which a producer can consume a producer value are also periodic.
- the scheduling zone can be determined statically for the consumer task p_0 as follows:

  1) Determine the intervals for every producer task p_i. They are given as $[kT+\varepsilon_i, kT+r_i]$.

  2) The scheduling zone, $[l_0, r_0]$, of p_0 is the intersection of the above intervals. More formally, it is given $[l_0, r_0] = \bigcap_{j\geq 1,\ p_j} [kT+\varepsilon_j, kT+r_j]$.

Note that $r_0 - l_0 \leq T$. Moreover, the computation of the task must be completed within the scheduling zone. This imposes the condition for the relative phasing of the consumer task that should be $\varepsilon_0 \in [l_0, r_0 - c_0]$. If the scheduling zone is the null set, then p_0 cannot be scheduled respecting the absolute temporal consistency constraint and this should be avoided. On the other extreme, if each value can be consumed anywhere within a period, then $r_0 - l_0 = T$.

Once the scheduling zones are determined for all consumer tasks, then the static scheduling problem consists of simply finding a schedule such that each consumer task can be scheduled within its scheduling zone. Three different cases have to be distinguished as follows:

Case 1: In the first case, the scheduling zones of all consumer tasks $p_0^1 \in P_1, p_0^2 \in P_2, ..., p_0^k \in P_k$ are equal to the period time. Therefore, the consumer tasks can be scheduled anywhere within period. A schedule can be constructed if $\sum_{i=1}^{k} p_0^i \leq T$. This is also the minimum condition for constructing a schedule for any set of periodic tasks.

Case 2: In this case, the scheduling zones of all consumer tasks $p_0^1 \in P_1, p_0^2 \in P_2, ..., p_0^k \in P_k$ are equal (denoted as $[r, l]$) but less than the period time. Therefore, the consumer tasks can be scheduled anywhere within a given interval. The minimum condition $\sum_{i=1}^{k} p_0^i \leq T$ is not sufficient. A valid schedule can be constructed if $\sum_{i=1}^{k} p_0^i \leq r - l$.

Case 3: When the scheduling zones are not the same for all consumer tasks, then the problem is not trivial anymore. The problem is equivalent to the scheduling problem with release times and deadlines, where l_0^1 is the release time and r_0^1 is the deadline of a consumer task p_0^1. The general case of this problem is known to be NP-complete [16] and therefore, computationally intractable. For special cases, where the computation time of each task is the same, then polynomial solutions are presented in [17]. Several heuristic solutions are presented in [18].

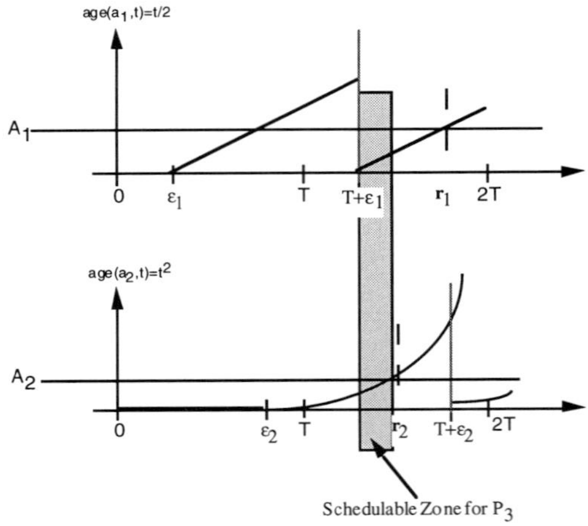

Fig.3. Scheduling zone for a subset of three tasks

<u>Special Cases</u>

When the age and the absolute threshold of all process variables are the same, that is, for all t, $age(a_i,t)=age(a_j,t)$ and $A_i = A_j$, we have either case 1 or 2 (depending on A_i). In this case, all tasks are synchronized, and their relative phasings are the same.

Another interesting case occurs when the absolute threshold is greater than the period time for all tasks, i.e. $A_i \geq T$. The result is case 1 and the tasks can be scheduled anywhere within a period time.

3.3. Dynamic Scheduling Problem (for unforeseen events)

As already mentioned in Section 1, the dynamic scheduling problem arises because the exact arrival times of the events are not known in advance. However, whenever such events arrive, the control system has to react to them within specified deadlines. The absolute threshold of an event defines the deadline within which control decisions have to be taken.

Let us first see what the occurrence of an event implies for a scheduler. When an event occurs, it invokes additional computation. We shall refer to the task that performs these computations as *aperiodic tasks.* From the scheduling point of view, the occurrence of an event is equivalent to adding additional tasks in a schedule. For this, a reactive scheduling method is well-suited because of the short response times. More specifically, this means that a reschedule of the static schedule has to be carried out whenever an event occurs.

Three relevant questions that arise are:

- when should the tasks be rescheduled ?
- what are the constraints for rescheduling ?
- how can the rescheduling be characterized ?

Rescheduling instants: As we consider only non-preemptive scheduling, a reschedule can be made only after an executing task completes its computation.

Scheduling constraints: Events are scheduled dynamically, and it should be done in such a way that the schedule of the periodic tasks are not affected (the reason for this is that, in many real-time systems, it is assumed that the periodic tasks should always meet their deadlines. On the other hand, missing deadlines of events once a while is tolerated. If an event cannot miss a deadline, then the task associated with the event can be replaced with an equivalent periodic task).

General measures: Examine the different cases discussed for static schedules. In all the three cases, the time available for adding tasks is the difference between the period time and the sum of the computation times of the periodic tasks, $T - \sum_{i=1}^{k} p_0^i$.

An interesting quantity is the size of a free block size (i.e. contagious time interval) available for events at a given instant of time. It can be used to evaluate the maximum number of aperiodic tasks that can be handled when the events arrive in a burst mode.

The maximum block size: In case 1 and 2, the maximum block size that can be achieved can be statically calculated. The maximum block size is equal to $2(T-\sum_{i=1}^{k} p_0^i)$. Unfortunately, in case 3, the maximum block size depends on the algorithm used. Moreover, calculating the maximum block size itself is a non-trivial problem.

In order to handle the events, the dynamic algorithm consists of two steps:

Step 1: find the best way in which the deadlines of periodic tasks can be preserved and blocks can be created in such a way that the maximum number of aperiodic tasks can be made.

Step 2: find a strategy such that the aperiodic task set can be scheduled in the most efficient way.

Step 1 depends on the complexity of the static algorithm. Whereas for step 2, we need criteria for defining a strategy. A few criteria are for example to maximize the number of aperiodic tasks that meet their deadlines, or minimize the penalty function of the tasks that miss their deadlines. In short, the general case of the problems can be complex and even NP-complete [16]. Thus, we note that in reactive dynamic scheduling of aperiodic tasks with absolute consistency constraints give rise to non-trivial problems.

The above steps are now illustrated with an example. The system is configured as follows:

- $P=\{p_1, p_2, p_3, p_4, p_5\}$, $E=\{e_1, e_2, e_3, e_4\}$.
- the computation times of p_4 and p_5 are c_4 and c_5 respectively.
- their data relationship is shown in Fig.4.
- the task subsets are $P'=\{(p_4, p_1, p_2), (p_5, p_1, p_2, p_3)\}$.
- the age functions of the tasks p_1, p_2, p_3 are the same and they are all assumed to have the same absolute threshold.

Therefore, we have case 2 of the static algorithm, where all the periodic tasks have to be scheduled within the same interval $[r,l]$, with $r-l<T$.

The behavior of the dynamic algorithm is shown in Fig.5. On the top, the maximum achievable block size is shown. It is equal to $2(T-(c_4+c_5))$. In the middle, at instant I_1 no events are registered and therefore, p_4 is scheduled. At the next interesting instant I_2, an event e_1 is registered. The maximum available block size at I_2 is $r-l-(c_4+c_5)$, because p_5

has to respect the deadline r. As e_1 can be accommodated in this block, e_1 is scheduled. Finally, at the bottom, as no events are registered at instant I_3, p_5 can be scheduled. At instant I_4 is $2(T-(c_4+c_5))-c(e_1)$ ($c(e_1)$ denotes the computation time of the aperiodic task associated with the event e_1).

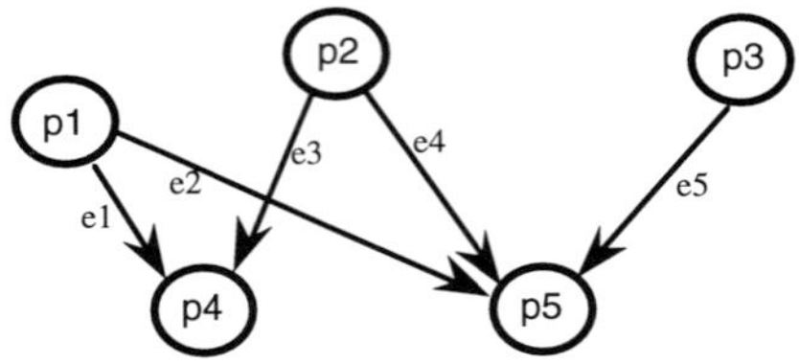

Fig. 4. Configuration of five tasks

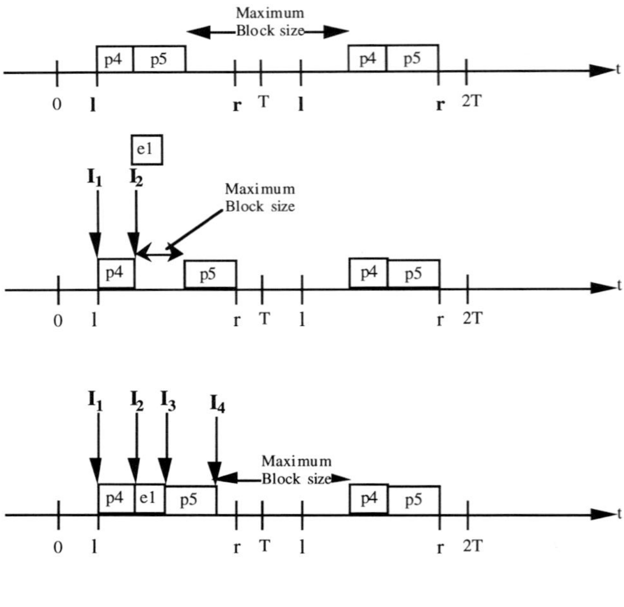

Fig.5. Different schedules of the dynamic algorithm

<u>Practical Problems</u>

The complexity of the dynamic algorithm is augmented if at a reschedule instant, two or more events have to be scheduled, then some criteria are required for choosing the appropriate one from the set of aperiodic tasks. Due to the enhanced complexity of the dynamic algorithms (the resulting problem is the combination of two NP-complete problems in the worst-case), it is difficult to find efficient solutions to the above problems that can be run in real time. In real-time systems, often, the dynamic algorithm has to take decisions and reschedule within few milliseconds.

Therefore, the main difficulty that arises with the scheduling is the limited time available for the reactive scheduling itself. Especially for dynamic reactive scheduling, different algorithms or heuristics must be evaluated, for which criteria are required. These can serve as a base for comparing different heuristics or solutions. To our knowledge, traditionally, the comparison is based on complexity. For example, one wants to know how many steps a method requires to achieve or approximate on optimal solution. In the presence of real-time constraints, a better question would be: "At the end of n time units, which algorithm will deliver the best solution ?". Practically, one could define an interval of time within which different algorithms are executed and compare the results. For a given problem, the *best* solution, achieved in this way, may not necessarily be the same as the optimal one in the classical sense. In other words, optimization in the presence of real-time constraints must be investigated.

4. CONCLUSIONS

In this paper, we addressed temporal consistency issues for a real-time system. In particular, we concentrated on the absolute consistency and introduced a function called *age*. The age function contains knowledge about the rate at which a process variable evolves. The implications of the age function and some applications of this function were also discussed.

The second part was dedicated to a scheduling problem, in the presence of absolute temporal consistency constraints. The scheduling problem was first handled for periodic systems. When combined with aperiodic systems (associated with the occurrence of events), it was shown that reactive scheduling methods can be used by a dynamic algorithm. The problems that arise when such techniques are used in the presence of real-time constraints was briefly discussed.

REFERENCES

[1] K. J. Astrom and B. Wittenmark, "Computer-controlled Systems: Theory and Design", 2nd Edition, Prentice-Hall International, Inc. 1990.

[2] Y. C. Ho (Ed.), "Discrete Event Dynamic Systems: Analyzing Complexity and Performance in the Modern World", Porceedings of the IEEE, Vol. 77, 1, 1989.

[3] P. Raja and G. Noubir, "A Functional Approach for Expressing Temporal Properties of Inputs and Outputs of Real-Time Applications", to appear in IEEE Workshop on Future Trends in Distributed Computing Systems, Lisboa, Portugal, Sep.22-24, 1993.

[4] P. Raja et al, "Characterizing Temporal Consistency in a Fieldbus Environment", in Singapore International Conference on Intelligent Control and Instrumentation, Singapore, Feb. 1992, pp. 766-771.

[5] P. Raja et al, "A Software Architecture for Maintaining Temporal Consistency in a Distributed Real-Time Environment", to appear in in 17th Annual International Computer Software and Applications Conference, Phoenix, Arizona, USA, IEEE Comp. Soc. Press, Nov 1993.

[6] P. Raja, "Temporal Consistency in Distributed Control Systems", IASTED Int. Conf. Robotics and Manufacturing, IEEE Robotics and Automation Soc., Oxford, England, Sep. 1993.

[7] H. Kopetz and K. H. Kim, "Consistency Constraints in Distributed Real-Time Systems", in M.G.Rodd and T.L. d'Epinay eds., Distributed Computer Control Systems 1988, Pergamon Press, 1988.

[8] H. Kopetz and K. H. Kim, "Temporal Uncertainties in Interactions among Real-Time Objects", in Ninth Symposium on Reliable Distributed Systems, Huntsville, Alabama, IEEE Comp. Soc. Press, Oct. 1990.

[9] X. Song and J. W. S. Liu, "Performance of multiversion concurrency control algorithms in maintaining temporal consistency", in 14th Annual International Computer Software and Applications Conference, Chicago, IL, USA, IEEE Comp. Soc. Press, Oct.31-Nov. 2, 1990.

[10] S. V. Vrbsky and K. Lin, "Recovering Imprecise Transactions with Rael-Time Constraints", Proc. IEEE 7th Symposium on Reliable Distributed Systems, pp. 185-193, October 10-12, 1988.

[11] R.J. Stokey, "AI Factory Scheduling: Multiple Problem Formulations", SIGART Newsletter, Oct. 1989, no.110, pp. 27-29.

[12] C. Pape and S.F. Smith, "Management of Temporal Constraints for Factory Scheduling", Technical Report, CMU-RI-TR-87-13, Computer Science Department, CMU, 1987.

[13] S.F. Smith, "Exploiting Temporal Knowledge to Organize Constraints", Technical Report, CMU-RI-TR-83-12, Computer Science Department, CMU, 1983.

[14] R.E. Korf, "Real-Time Heuristic Search First Results", Proc. AAAI Conference, 1987.

[15] M.S. Fox, "Constraint-directed search: A Case Study of Job-Shop Scheduling", Technical Report, CMU-RI-83-22, Computer Science Department, CMU, 1983.

[16] M.R. Garey and D.S. Johnson, "Computers and Intractability: A Guide to the Theory of NP-Completeness", W. H. Freeman and Company, New York, 1979.

[17] B. Simons, "A Fast Algorithm for Single Processor Scheduling", Proc. 19th Annual Symposium on Foundations of Computer Science, IEEE Computer Society, Long Beach, CA, pp. 246-252.

[18] D. Gunsfeld, "Bounds for Naive Multiple Machine Scheduling with Release Times and Deadlines", Journal of Algorithms, vol.5, 1984.

Knowledge-Based Reactive Scheduling (B-15)
E. Szelke and R.M. Kerr (Editors)
Elsevier Science B.V. (North-Holland)

Configurable Systems for Reactive Production Management

Stephen F. Smith and Ora Lassila

Center for Integrated Manufacturing Decision Systems
The Robotics Institute
Carnegie Mellon University
Pittsburgh, PA 15213 USA

In this paper, we outline current work into the development of *reconfigurable scheduling systems*: systems that enable rapid customization to specific production environments through encapsulation, extension, and reuse of component scheduling "services" (e.g., domain modeling primitives, constraint management and analysis techniques, solution subprocedures, problem decomposition and configuration heuristics). Our approach is grounded on two basic premises: (1) that system organization and decision-support "services" should directly reflect the inherently reactive nature of decision-making in practical scheduling environments, and (2) that diversity in the character and requirements of different application environments will invariably require different specialized scheduling support services. We adopt a general constraint-based model of scheduling as an iterative, opportunistic process of schedule revision, which provides an architectural framework for formulating the configuration problem, and utilize object programming techniques to compositionally construct component services. Our overall goal is the development of an application building environment, which combines a "tool box" of basic modeling and scheduling primitives with facilities for assembling, aggregating and specializing these primitives to define the decision support functionality (or services) required in a given application context. As new services are composed, they can be encapsulated as additional, higher level tools and are available for reuse in subsequent applications.

Keyword Codes: D.1.5; I.2.1
Keywords: Programming Techniques, Object-oriented Programming; Artificial Intelligence, Applications and Expert Systems

1. INTRODUCTION

Two broad observations can be made with respect to the scheduling problems that must be faced in practical production environments:

- They can rarely be treated strictly as optimization problems. To be sure, optimizing capabilities are important, but the overarching goal of producing a good manufacturing system *behavior* requires larger attention to the ongoing scheduling process.

Schedule construction in practice tends to be an dynamic reactive process, involving negotiation among decision-makers with different organizational responsibilities and goals (e.g., marketing, engineering, production). An initial schedule is built, problematic or unsatisfactory aspects of the result are identified, requirements are relaxed or strengthened, schedule modifications are made and so on. In this context, the current schedule serves as an important nominal reference for identifying and specifying changes. While the focus is on improving the acceptability/quality of the solution, considerable pragmatic value is placed in retaining continuity in the solutions produced across iterations. Likewise, as unexpected events occur in the execution environment, it is important to preserve continuity in domain activity while making those schedule changes necessary to insure continued feasibility and attendance to overall performance objectives. Both of these aspects of the scheduling process place a premium on incremental, reactive scheduling capabilities.

- There is considerable diversity in the character of the scheduling problems faced in different manufacturing environments. While commonality can be found in overall solution and decision-support requirements, the structure of the manufacturing system, the types of constraints that dominate the problem, the types of uncertainties that must be confronted and the desired scheduling objectives vary considerably from one environment to the next. The utility of a given scheduling technology will directly depend on the level of fidelity of its underlying model with the operating constraints and conditions of the target manufacturing environment. Accordingly, we can expect that the heuristics and solution procedures required for effective decision-support will likewise vary from application to application.

The requirements implied by these observations are at odds with the decision-support capabilities provided by most contemporary production management tools. Existing systems generally operate with respect to simplified models of domain constraints, limiting their relevance and utility to operational decision-making. Moreover, existing systems are more often than not based on the use of inflexible schedule generation procedures (e.g., back-end optimizers or simulators). Such approaches force the user into an indirect and inefficient reactive decision-making cycle (where the user must hypothesize parameter changes that will bias the procedure toward the desired solution, run the procedure, interpret the results and iterate), and such approaches provide no support for maintaining stability in the solutions produced over time.

Recent research in knowledge-based scheduling has made progress toward overcoming these inadequacies [22]. The advantage of heuristic scheduling procedures that are directly based on knowledge-rich models of the actual operating constraints and objectives of the target environment has been demonstrated in several application domains. Similarly, recent development of techniques and heuristics for incremental schedule revision (e.g., [3, 4, 16–18, 20, 21, 26] has provided scheduling functionality that more closely parallels the inherently reactive nature of the scheduling process.

Despite the promise of these results, their impact in operational production management settings remains low. Somewhat ironically, the source of strength of knowledge-based reactive scheduling techniques in overcoming the inadequacies of current production management technologies - incorporation of knowledge about the constraints and objectives of the particular manufacturing environment - also complicates the application building process. As observed above, different production environments invariably present different scheduling challenges and complexities, and efforts to apply knowledge-based reactive scheduling technology to specific operational environments largely remain time-consuming, "one-off" design and development projects.

In this paper, we describe work aimed at simplifying this application building process. Starting from a flexible and general model of reactive scheduling as an iterative, constraint-directed schedule manipulation process, we focus on the development of a system architecture that promotes appropriate instantiation of this model in specific manufacturing environments. In contrast to the perspective of a universally applicable heuristic search framework that can be programmed with the evaluation criteria of a given application (e.g., [6, 26]), our approach presumes that different applications will typically require different (more specialized) search and solution procedures, and it focuses on providing support for this solution configuration process. Our overall goal is to develop a *reconfigurable scheduling system*, an application building environment which combines a "tool box" of basic modeling and scheduling primitives with facilities for assembling, aggregating and specializing these primitives to configure the decision support functionality (or services) required in a given application context. As new services are composed, they can be encapsulated as additional, higher level tools and are available for reuse in subsequent applications.

2. CONSTRAINT-BASED SCHEDULING MODELS

In arguing against universal solution procedures, we are not suggesting that there is no commonality in the scheduling functionality required in different application domains. In fact, we believe quite the opposite; that required solution procedures in various applications can be seen as more or less similar if they are viewed compositionally, and exploitation of this fact is the key to achieving broad applicability. To do so, however, first requires that one settle on an appropriate scheduling model.

Constraint-based frameworks provide a model well-suited to the reactive decision-making requirements of practical scheduling domains. In broadest generality, this model defines a problem solving organization that distinguishes two components: a *decision-making* component, responsible for making choices among alternative scheduling decisions and retracting those that have since proved undesirable, and a *constraint management* component, whose role is to propagate the consequences of decisions and incrementally maintain a representation of the current set of feasible solutions (detecting inconsistent solution states when they arise). Schedule construction, revision, and improvement proceed iteratively within a basic *decide and commit* cycle. Most project management tools and several interactive scheduling systems [7, 15] are direct implementations of this model, with the user as the decision-making component.

Of more interest and importance in most scheduling environments, are extensions of this basic model that off-load more decision-making responsibility to the system. The "spreadsheet" style of interaction provides a natural framework for "what-if" experimentation and iterative solution development, but interaction and decision-support is typically required at a much more aggregate decision-making granularity. One extension to the basic constraint-based model that preserves the "direct manipulation" style of interaction is one that refines the decide step of the cycle into a two-step process of formulating and executing decision-making actions. Within this model, *action formulation* is concerned with isolating a particular subproblem and *action execution* results in solution of this subproblem. The user, who interacts with system processes by formulating actions, is able to manipulate solutions in terms of higher-level (and more comprehensible) task-oriented perspectives (e.g., reschedule job x to start tomorrow), with the system's subproblem solution procedures providing, from the user's viewpoint, an amplification of deductive constraint management functionality; a more sophisticated (and typically heuristic) "propagation of effects".

In the simplest case, there is a direct mapping between user specifiable actions and system solution procedures, in which case the user holds complete responsibility for action formulation. The COMPASS interactive scheduling framework [5], for example, is organized in this fashion. In our view, however, the user should be able to operate in terms of much more ill-structured action specifications (e.g., relax weekend capacity constraints and reschedule late orders while minimizing overtime personnel, resolve the conflicts introduced into the schedule by the breakdown of machine x). This, more flexible viewpoint implies that the system must participate actively in structuring the appropriate subproblem to solve (e.g., in determining the appropriate scope of change, in translating objectives and preferences into appropriate heuristic revision procedures), and that subproblem solution may require coordinated execution of several solution procedures.

3. THE RECONFIGURABILITY PROBLEM

Our perspective on the development of reconfigurable scheduling systems derives from previous work with the OPIS scheduler [20, 23], which advocated a similar (but narrower) perspective toward reactive scheduling system development. OPIS implements a framework for incremental, reactive scheduling based on the use of a set of solution subprocedures with differential optimization and conflict resolution capabilities, which are dynamically selected from and applied to best respond to current (re)scheduling needs and opportunities. While the principal emphasis in this research was on mechanisms and heuristics for integrating methods to balance scheduling, schedule stability and system responsiveness objectives, this opportunistic, multi-perspective scheduling methodology, as well as the supporting architecture developed for specification and coordination of system activity, are directly relevant to the larger problem of configuring scheduling services for various application domains.

At the same time, we can retrospectively identify several design deficiencies in the OPIS scheduler from the standpoint of reconfigurable scheduling systems:

- Inflexibility in solution representation/constraint management assumptions - In the original design of the OPIS constraint management subsystem [13], specific assumptions as to solution representation were adopted and specialized constraint propagation and conflict detection mechanisms were developed to manage these representations in response to solution changes; in essence, generality in the types of constraints that could be represented and managed was traded off to achieve sufficient propagation efficiency to accommodate manipulation of full-scale production schedules. This functionality has proven quite satisfactory across a number of distinct production scheduling application domains. However, more recent work in adapting the original manufacturing production scheduler to the domain of crisis-action deployment scheduling (admittedly a quite different problem domain with quite different solution management requirements), necessitated fairly significant extensions to this constraint management infra-structure, which in turn resulted in a rippling of changes to other dependent system components (e.g., conflict analysis routines). Interestingly, the system's original solution subprocedures proved quite effective once these infra-structure adaptations were made, yielding better solutions than existing transportation scheduling tools in a comparative performance study [24]. The point here is not that specializing kernel system functionality such as constraint propagation is a bad idea; specialization is often crucial to achieving sufficient system efficiency. The problem lies in appropriately organizing system functionality so that component services (e.g., solution representations, constraint management techniques) appropriate to the domain under consideration can be easily substituted and configured. In this regard, the recent work of LePape is a strong step in the right direction [14].

- Overcommitment in the control infra-structure - Reactive scheduling in OPIS is based on a fairly general process of matching current rescheduling needs and opportunities (as inferred through analysis of the characteristics of the current solution state) to the differential capabilities of constituent revision subprocedures and heuristics. However, in providing a structure for specifying and coordinating reactive scheduling strategies, the supporting OPIS control architecture makes several specific assumptions as to the mechanics of this process (e.g., how external events are mapped to solution structures, how problems are detected, how the focus and scope of revision tasks are determined, how the ripple effects of local revision actions are accounted for). Many aspects of this architecture reflect the original system design orientation toward systems for incremental, reactive response to unexpected executional circumstances. However, when viewed alternatively from the larger perspective of supporting response to ill-structured user formulated actions, these architectural commitments are too strong. In fact, as was argued above with respect to underlying solution representations and constraint management techniques, we expect that user requirements in different application environments will imply different levels of architectural commitment. Reconfigurability requires a system design that retains the flexibility to adapt the mechanics of the control cycle to best fit the user requirements of specific domains.

Both of these deficiencies point to one overarching problem: the system design does not properly anticipate the potential necessity for change/extensibility at any level of system functionality. Configuration flexibility is only provided with respect to pre-identified functional components (e.g., the addition/substitution of alternative solution subprocedures and alternative parameterizations of these subprocedures, specialization of domain descriptions to encode idiosyncratic application constraints), and the control infra-structure, while clearly relevant to an iterative, opportunistic model of user-system interaction, is too confining in its assumptions.

4. TOWARDS RECONFIGURABLE SYSTEMS

The shortcomings of the OPIS scheduler identified in the previous section highlight the need for stronger emphasis on configurability and extensibility in scheduling system design. It is very difficult to anticipate all future needs for system extension; instead, a technique must be used that allows specialization, modification and extension of *any* component of the system. Within OPIS, *frame-based* representation techniques were used to provide both a repository of primitives for modeling domain constraints and a framework for specifying strategic (i.e. method configuration and parameterization) control knowledge. While such representational formalisms provide the structure to define flexible and expressive scheduling models, they provide no explicit mechanisms for encapsulation and information hiding. Pragmatically, this greatly complicates specification of, and adherence to, a layered model semantics, which is essential to the development of reconfigurable and extensible software systems[1]. As has been previously observed [8, 11, 12], modern *object-oriented* programming technologies provide a much more direct and effective approach to specifying model semantics, through explicitly defined protocols for interaction with model components.

Our approach to simplifying knowledge-based scheduler construction relies heavily on object-oriented programming techniques and software reuse, allowing schedulers and other production planning applications to be constructed as a "differential" process, focusing primarily on the differences between existing software and the system being constructed. Object-oriented programming techniques can provide high reusability of software, but only if the system design project places special emphasis on the design of reusable *components* (e.g., [25]). The design of these components must be carried out with generality and extensibility in mind.

A scheduler construction project (potentially as part of a larger production management system design) begins with:

- Information gathering and *knowledge acquisition*, aiming to provide the system designer with a sufficient amount of detailed information about the production system.
- *Modeling* and *analysis* of the production system (e.g. production plant and process, logistics system).

[1] In particular, if all the slots of a frame are accessible, no encapsulation is achieved: this results in fragile implementations that are difficult to modify and maintain.

In an object-oriented approach to software construction, these steps constitute an object-oriented analysis of the scheduling system (they also constitute part of the design). The approach taken in our reconfigurable framework is the introduction of a common scheduling and production planning ontology which serves as the starting point for a more detailed analysis of the target system. The system offers the scheduling system designer a class library of common and general scheduling concepts, such as *activities*, *resources*, *products* and *orders*. Constructing a scheduler or some other production planning application using our approach consists of the following:

- *Selecting* suitable classes from the library, matching features of the target system with those of the library.
- *Combining* the selected classes into more complex services, using both conceptual (i.e. multiple inheritance) and structural (i.e. aggregation and delegation) techniques.
- *Extending* the existing classes to provide domain-specific functionality when necessary. Typically this is done by specializing or overriding methods provided by the library.

The basic class library provides a general scheduling ontology. This ontology can be specialized for specific fields of production (for example, we have built a transportation domain ontology, allowing us to easily build transportation-related applications). The general and the field-specific ontologies can then be used to build company-specific ontologies and actual production planning applications. It can be observed that although our system focuses on production planning and scheduling, this approach is in many ways similar to that of the CIM-OSA -architecture [9] (c.f. its Generic Level, Partial Models and Particular Models).

The general philosophy of application construction is to use the techniques described above to build increasingly complex (and specialized) services, ultimately resulting in an application. The classes designed and specialized in this process can be reused in subsequent applications. An application designer may accumulate a library of specialized classes, making his task easier as the library grows.

4.1. Library Design

Different schools of thought exist as to how the object-oriented design process should be structured (e.g., [1, 25]). Mainly these differ just in details, the overall process consisting of two main steps: First, the proper *concepts* have to be identified, clearly specifying the structural and behavioral roles of each concept. Second, a *protocol* of interaction between different classes has to be identified and specified. In the design of a class library, a protocol consists of a set of methods which are chosen to allow system extension. In other words, the protocol must be as general as possible, without compromising performance.

By specifying and documenting the protocol of interaction between different objects, the implementation of the class library (and thus the implementation of the scheduler) is "opened up" (allowing client programmers to extend it as they see fit), thus departing from the traditional "black box" mentality to modular software, and allowing limited visibility inside different modules. This approach is similar to the *metaobject protocol* implementation techniques of CLOS [10], where the objects of the implementation (also called *metaobjects*) allow the behavior of the system to be controlled and modified. If we think of traditional software reuse (e.g., [2]), this goes beyond the simple function library approach, thus complicating liability issues, for example. These issues, however, are beyond the scope of this paper.

The classes in our library have various roles and functions. A large number of the classes is intended for modeling production systems, but will also function as providers of important services in a scheduling application. Some classes provide control functions, and serve as the building blocks of control architectures. Finally, instances of some classes are "pure" metaobjects: their role is to provide certain services (e.g. time and calendar services, error handling) and exist as vehicles for system extension (the client programmer can substitute them with objects from his own classes to provide modified or extended functionality). In general, most classes can be roughly divided into three main categories:

- *Base Classes* are (usually) not instantiable; they define protocols and provide base and default functionality of a particular concept. They are easy to understand and use, since they each implement a limited, reasonably well defined set of functionality. Complex concepts have been broken into several classes. Examples of base classes are `resource`, `operation` and `conflict`.
- *Specialized Classes* can be instantiated, i.e. used without specialization. They typically refine and extend the protocols of the base classes. Many classes in the field-specific ontologies belong to this category. Examples of specialized classes: `batch-resource`, `airport`, `resource-unavailability-conflict`.
- *Mixin Classes* cannot be used alone, they must always be "mixed" with base and specialized classes. They typically implement functionality, completely or nearly orthogonal to the base classes, or provide simple refinements to the base behavior. Examples of mixin classes are `movable-resource-mixin` (to provide location tracking) and `min-max-duration-mixin` (to allow storage of lower and upper bounds of duration).

Currently our core library has 80 classes available for the client programmer. We have implemented a specific model for military transportation planning which has 40 classes, all specializations of the core classes.

4.2. Examples

The use and nature of the class library is best demonstrated through examples of how to select, combine and specialize classes and how to build services and applications. It is imperative to understand that building systems using a library like this is a "differential" process: The library provides concepts and mechanisms common to production planning applications, and to construct an application only differing concepts and behavior have to be defined and implemented. Through reusability of the library components we are able to construct applications rapidly and reliably.

As an example of an application, we will discuss a module implemented for a larger planning application. The purpose of the module is to perform feasibility checks on plans generated by human planners using the planning application. Both temporal feasibility as well as resource availability are verified. This system roughly consists of three parts:

- An *input module*, responsible for parsing the input from the planning application and instantiating a model of the plan. Three different things take place when a plan is instantiated: appropriate resource objects are created, activities are created (naming the resources that execute these activities) and the activities are linked using temporal relations. All objects are created either from the core library classes directly or from classes that have been specialized from the class library.
- A *checker module*, responsible for checking the feasibility of the plan. This module employs the *Time Bound Propagator*, one of the services provided by the class library, to accomplish two things: to detect any cycles in the activity graph, and to establish temporal bounds for each activity. The kernel of the propagator is a constraint path-consistency algorithm capable of propagating temporal constraints in a constraint graph consisting of activities and temporal relations.
- An *output module*, responsible for gathering the results of the feasibility check and communicating them to the planning application. It collects the *conflicts* created by the propagator, and filters these to provide a description of the potential infeasibilities. It then translates both the conflicts and the time bounds of all activities into an appropriate format.

Activities in the plan have both a maximum and a minimum duration, both of which are not necessarily known. The next example shows how we define our own activity class, based on the base class `operation` and the mixin class `min-max-duration-mixin`:

```
(defclass plan-activity (operation min-max-duration-mixin)
  ()
  (:default-initargs
    :setup-duration 0))
```

The resulting class `plan-activity` behaves like the basic operation class, but is capable of storing (and interpreting) both a minimum and a maximum duration. Here the specialization has been effected by selecting the appropriate base classes and combining them through multiple inheritance to provide a problem-specific class. More complex specialization can be achieved by extending the behavior of the objects. This is done by overriding or specializing methods of the objects' protocols.

Let us consider scheduling material movement in the manufacturing domain. Movement activities typically have a duration that depends on the distance covered and the speed of the AGV resource. This can be implemented by overriding the default duration methods of the `operation` class. To illustrate this, we will give the definitions for a simple transportation domain:

```
(defclass movement-by-agv (operation)
  ((origin
     :initarg :origin
     :reader operation-origin)
   (destination
     :initarg :destination
     :reader operation-destination)))

(defclass agv (batch-resource)
  ((velocity                ; expressed in miles per hour
     :initarg :velocity
     :reader resource-velocity)))

(defmethod operation-compute-run-duration ((op movement-by-agv)
                                           (res agv)
                                           &optional type)
  (declare (ignore type))
  (* (/ (location-distance-between (operation-origin op)
                                   (operation-destination op))
        (resource-velocity res))
     (/ (time-units-per-day) 24)))
```

The class library provides a base class, `location`, which can be used for objects of the *origin* and *destination*. The protocol also specifies a method for locations, called `location-distance-between`, this can be specialized in different location classes (for example, locations on the factory floor, seaports, cities, etc.)[2]. This example is no doubt somewhat simplified but provides an insight into how specialization can take place.

Our last example should illustrate the mechanics of a lower-level protocol: A schedule is stored in *resource* objects by maintaining a representation of the available capacity of each resource over time. A protocol exists for querying this representation, allowing other program modules to find out about available capacity and windows of time where scheduling can take place (for example, the Time Bound Propagator relies heavily on this protocol). Each *operation* class has a method which queries the time available for the particular operation to be scheduled. Since operations know about their required resources, the method will call a resource method for querying time blocks. This method has several definitions, depending on the type of resource (for example, an "atomic" resource only allows one

[2]The function `time-units-per-day` used in the example is part of the time and calendar services of the library.

operation to be performed on it at a time, a "batch" resource allows time-synchronized execution of several operations, an "aggregate" resource represents a collection of resources, each capable of independently executing operations; in each case the available capacity is treated differently). A callback to the client is provided through the use of *time block strategies* which are objects that ultimately decide about time blocks found. The resource time block method provides this by calling `strategy-enough-capacity-p` and `strategy-check-blocks`. These methods allow, for example, capacity constraints to be relaxed, in order to resolve overconstrained situations.

To allow a scheduler to hypothesize about, and schedule into, infinite capacity, the predicate method `strategy-enough-capacity-p` may be overridden. This method determines whether the available capacity is large enough to allow allocation of the required capacity. In order to create the "illusion" of infinite available capacity, we may redefine this method to return true even in cases when there is not enough capacity available. To instruct the use of time intervals during which minimal overallocation takes place, the method `strategy-check-blocks` may be redefined: its role is to inspect time blocks and decide which ones are preferred.

The last example can be extended to implement a reactive scheduling technique where formerly allocated operations can be "bumped", i.e. the capacity they are reserving can be regarded free. To implement this, we need to assign priorities to each operation (or corresponding production plan), define a new strategy (say, `bumping-time-block-strategy`) and specialize the method `strategy-enough-capacity-p`. This method would, in case of inadequate amount of available capacity, find operations of lower priority whose capacity could be considered available for allocation.

4.3. Development Framework

The classes and protocols described in the previous examples are part of a larger framework, the *configurable scheduling framework*, which establishes a full hierarchy of protocols implementing the model of constraint-based scheduling developed in earlier sections. The framework provides a starting point for a scheduling application builder, who will replace abstract classes of the framework with more specialized classes that suit the problem at hand. This approach is analogous to modern micro-computer application development frameworks, which provide the basic functionality of a complex user-interface in the form of an application "skeleton" (e.g., [19]). In our case, the "skeleton" is an empty, generic constraint-based scheduling system. Solutions and classes defined in the framework will probably suit the majority of application needs. However, there is always the possibility to replace any component of the system with a different or more specialized component. This flexibility is directly attributable to the abstract layering of object interaction protocols provided by the framework.

To replace a certain component in the abstract framework, it is necessary to (1) decide which protocols the new component will commit to, and (2) define or specialize an appropriate new class. Considering the structure where primitive classes and services are combined into increasingly complex services, there are two choices in replacing a component:

- The component commits to both upper and lower level protocols; in this case only the component itself needs to be replaced. This is possible since the internal protocols of the framework are documented.
- The component only commits to the upper level protocol; in this situation the lower levels of the service are either implemented from scratch or possibly with the help of some of the existing lower level services.

By providing a "complete" scheduling system framework we provide a structural backbone for scheduling system (re)configuration. Application building becomes a differential process, since all the basic services already exist, and only differences to the target scheduler need to be addressed. Of course, decisions also need to be made concerning interfaces to other systems (e.g. database interfaces, user interaction). We plan to extend the framework to provide predefined modules for these purposes in the future.

5. CONCLUSIONS

In this paper, we have advocated the development of reconfigurable scheduling systems as a means of simplifying the application building process in different production environments. To provide a structure for reconfiguration, we have adopted the constraint-based scheduling model as a likely candidate for the architecture of future scheduling systems. As pointed out, constraint-based techniques support the development of intelligent, localized rescheduling capabilities, thus providing an intuitive base for direct-manipulation, "spreadsheet-like" user interaction. Likewise, the basic division of problem solving architecture into decision making and constraint management components is compatible with the view of a scheduler as a decision support system rather than a fully automatic component of a production management system.

Reconfigurability can be seen as a means to rapid development of reliable scheduling systems and other intelligent production planning applications, and can also be viewed as a source of flexibility from the end user standpoint. The object-oriented techniques outlined in the previous section will serve as an enabling technology for reconfigurability, by transforming application development into a differential and incremental process, promoting software reuse, and ensuring unlimited extensibility.

REFERENCES

1. Booch, Grady, "Object-Oriented Design with Applications", Benjamin/Cummings, Redwood City, CA, 1991.
2. Bott, M.F. and P.J.L.Wallis, "Ada and Software Reuse", in *Software Reuse with ADA*, (eds. R.J. Gautier and P.J.L. Wallis), IEE, 1990.
3. Burke, P., and P. Prosser, "A Distributed Asynchronous System for Predictive and Reactive Scheduling", Tech. Report AISL42, Dept. of Computer Science, University of Strathclyde, 1989.

4. Collinot, Anne and Claude Le Pape, "Adapting the Behavior of a Job-Shop Scheduling System", *International Journal for Decision Support Systems*, 7(3), pp 341-353, 1991.
5. "Compass 2.0 User's Guide", Planning and Scheduling Group, McDonnell Douglas Space Systems Co., December 1992.
6. Fox, Mark S. and Katia P. Sycara, "Overview of CORTES: A Constraint Based Approach to Production Planning, Scheduling and Control", in *Proceedings of the Fourth International Conference in Expert Systems in Production and Operations Management*, (ed. Martin D. Goslar), Management Science Department, College of Business Administration, University of South Carolina, 1990.
7. Elleby, P., H.E. Fargher, and T.R. Addis, "Reactive Constraint-Based Job-Shop Scheduling", in *Expert Systems and Intelligent Manufacturing*, M.D. Oliff (ed.), North-Holland, New York, 1988.
8. Hynynen, J. and O. Lassila, "On the Use of Object-Oriented Paradigm in a Distributed Problem Solver", *AI Communications* 2(3) 142-151, 1989.
9. Jorysz, H. R. and F. Vernadat, "Defining CIM Enterprise Requirements using CIM-OSA", in *Computer Applications in Production and Engineering: Integration Aspects (CAPE'91)*, (eds. G. Doumeingts, J. Browne and M. Tomljanovich), Elsevier Science Publishers B.V. (North-Holland), Amsterdam, 1991.
10. Kiczales, Gregor, Jim des Rivières and Daniel G. Bobrow, "The Art of the Metaobject Protocol", MIT Press, Cambridge (Mass.), 1991.
11. Lassila, O., "Frames or Objects, or Both?" *Workshop Notes from the Eight National Conference on Artificial Intelligence (AAAI-90): Object-Oriented Programming in AI.* Boston (Massachusetts, U.S.A.), July, 1990. [Also published as report HTKK-TKO-B67, Otaniemi (Finland), Department of Computer Science, Helsinki University of Technology, 1990].
12. Lassila, O., "The Design and Implementation of a Frame System", Master's Thesis, Otaniemi (Finland), Faculty of Technical Physics, Helsinki University of Technology, 1992.
13. LePape, C. and S.F. Smith, "Management of Temporal Constraints for Factory Scheduling", in *Proceedings IFIP TC 8/WG 8.1 Working Conference on on Temporal Aspects of Information Systems (TAIS 87)*, 1987.
14. Le Pape, C., "Using Object-Oriented Constraint Programming Tools to Implement Flexible 'Easy to Use' Scheduling Systems", in *Proceedings NSF Workshop on Intelligent Dynamic Scheduling for Manufacturing Systems*, Cocoa Beach, FL, January, 1993.
15. Meng, C.C. and M.Sullivan, "LOGOS: a Constraint-Directed Reasoning Shell for Operations Management", *IEEE Expert*, 6(1), 1991.
16. Miyashita, Kazuo and Katia Sycara, "Adaptive Case-Based Control of Schedule Revision", in *Intelligent Scheduling*, (eds. M. Fox and M. Zweben), Morgan Kaufmann Publishers, 1993.
17. Ow, P.S., S.F. Smith and A. Thiriez, "Reactive Plan Revision", *Proceedings AAAI-88*, St. Paul, MN, August, 1988.
18. Sadeh, Norman, "Micro-Opportunistic Scheduling: The Micro-BOSS Factory Scheduler", in *Intelligent Scheduling*, (eds. M. Fox and M. Zweben), Morgan Kaufmann Publishers, 1993.

19. Schmucker, Kurt. J, "Object-Oriented Programming for the Macintosh", Hayden Publishing Company, Hasbrock Heights, NJ, 1986.
20. Smith, S.F., P.S. Ow, N. Muscettola, J.Y. Potvin, and D. Matthys, "An Integrated Framework for Generating and Revising Factory Schedules", *Journal of the Operational Research Society*, 41(6), June, 1990.
21. Smith, S.F., N. Keng, and K. Kempf, "Exploiting Local Flexibility During Execution of Pre-Computed Schedules", in *Applications of AI in Manufacturing* (eds. D. Nau and C. Tong), MIT Press, July, 1992.
22. Smith, S.F., "Knowledge-Based Production Management: Approaches, Results, and Prospects", *Production Planning and Control*, 3(4), pp. 350-380, 1992.
23. Smith, S.F.,"OPIS: A Methodology and Architecture for Reactive Scheduling", in *Intelligent Scheduling*, (eds. M. Fox and M. Zweben), Morgan Kaufmann Publishers, 1993.
24. Smith, S.F. and K.P. Sycara, "A Constraint-Based Framework for Multi-Level Transportation Scheduling", CMU Robotics Institute Technical Report, June 1993.
25. Wirfs-Brock, Rebecca, Brian Wilkerson and Lauren Wiener, "Designing Object-Oriented Software", Prentice Hall, Englewood Cliffs, NJ, 1990.
26. Zweben, Monte, Eugene Davis and Michael Deale, "Iterative Repair for Scheduling and Rescheduling", Technical Report, NASA Ames Research Center, MS 244-17, Moffett Field, CA 94035.

ACKNOWLEDGEMENTS

The authors would like to thank the following people of fruitful discussions and advice during the design of the system described and during the preparation of this paper: Robert Aarts, Marcel Becker, Marcia Laurenson, Seppo Törmä.

Knowledge-Based Reactive Scheduling (B-15)
E. Szelke and R.M. Kerr (Editors)
Elsevier Science B.V. (North-Holland)

Adaptive Schedule Repair

Katia P. Sycara and Kazuo Miyashita

The Robotics Institute
School of Computer Science
Carnegie Mellon University
Pittsburgh, PA 15213
katia@cs.cmu.edu

Matsushita Electric Industrial Co. Ltd.
2-7 Matsuba-cho, Kadoma, Osaka 571, JAPAN.
miya@ei.sanken.osaka-u.ac.jp

Abstract

We describe a unified framework for job shop schedule optimization in predictive and reactive contexts by iterative revision. The approach elicits user context-dependent scheduling preferences in the form of cases, and re-uses them to dynamically guide schedule revision decisions. The methodology integrates case-based reasoning (CBR) mechanisms for incremental accumulation and re-use of past repair experiences that incorporate user preferences and schedule evaluation explanations, and constraint-based scheduling for propagation and resolution of the effects of repair. This framework is implemented in a system called CABINS. Experimental evaluation of CABINS as compared with other scheduling methods shows that the approach is able to (a) capture and effectively utilize user's scheduling preferences that were not present in the scheduling model, (b) outperforms other scheduling methods in both predictive schedule generation and reactive response to unpredictable execution time events along a variety of criteria that have been recognized as crucial in real operating environments.

Keyword Codes: I.2.1; I.2.6; I.2.8
Keywords: Artificial Intelligence, Applications and Expert Systems; Learning; Problem Solving, Control Methods, and Search

1. Introduction

Job shop scheduling deals with allocation of a limited set of resources to a number of activities (operations) associated with a set of jobs so as to respect given temporal relations and resource capacity restrictions in order to optimize a set of objectives, such as minimize tardiness, minimize work in process (WIP), maximize resource utilization etc. Scheduling is difficult to automate for a variety of reasons. First, it is an NP-complete problem [7]. Second, due to the tight interactions among scheduling constraints and the often conflicting nature of optimization criteria, it is impossible to assess with any precision the impact of a scheduling decision on the global satisfaction of optimization criteria. Third, the definition/evaluation itself of what is a "high quality" schedule is fraught with difficulties because of the need to balance conflicting objectives and trade-off among them. Such trade-offs typically reflect user preferences and the presence of additional domain constraints not captured in the scheduling model. The value of incorporating context-dependent user preferences in operational scheduling environments is becoming increasingly recognized (e.g., [12]). Finally, a schedule that is only *predictive* (i.e. it is created assuming that the world is static and predictable) will be brittle. Since operational environments (e.g. factories) are dynamic, it is clear that any effective scheduling system must be able to be *reactive*, i.e. perform schedule revision in response to unforeseen events during schedule execution.

Knowledge-based scheduling has been viewed as a constraint satisfaction problem (CSP). [6, 18, 20, 9, 23, 13, 21] Constraint satisfaction provide a strong foundation for generating feasible schedules efficiently [2, 11, 8, 19, 5]. But, basic constraint satisfaction search techniques do not give guidance in constraint relaxation decisions when a feasible schedule does not exist, neither do they guarantee finding optimal solutions. [22]

In this paper, we describe an approach that provides a unified framework for acquisition of user preferences, predictive optimal schedule generation based on these preferences, and reactive schedule management in response to unforeseen events. Our approach views both predictive and reactive schedule generation as an incremental schedule *revision process* that operates on a complete but "flawed" schedule. Revision-based approaches to predictive schedule generation have been adopted by (e.g., [25, 1, 13]). In [13], the MIN-CINFLICTS heuristic that chooses the repair that minimizes the number of remaining conflicts, was shown to perform well in solving N-queens CSP problems. However, unless the initial to-be-repaired schedule was judiciously crafted, MIN-CONFLICTS was shown to perform poorly in job shop scheduling problems in experiments reported in [16, 15]. In [25], simulated annealing is used as a repair method for predictive schedule generation in space shuttle ground processing. The algorithm sacrifices schedule quality to achieve efficiency and moreover is not guaranteed to produce conflict-free (executable) schedules. While this might not be a serious drawback in space shuttle processing where apparently humans can resolve residual conflicts by multiplexing resources among physically proximate tasks, the tight interactions in job shop scheduling make such fixing extremely difficult without further compromising schedule quality. In [1], chronologies are used to locate bottlenecks in oversubscribed spacecraft mission scheduling and associated heuristic metrics guide repair actions for bottleneck resolution. This technique is modeled after the repair behavior of human mission schedulers who identify and classify problem areas and use heuristic techniques to reduce resource contention. Unlike job shop scheduling where all jobs must be scheduled, in mission scheduling the requested tasks far exceed the capacity of spacecraft resources thus necessitating scheduling only a subset of requested tasks to maximize the total number of scientific experiments that can be performed. In [17], while predictive schedules are generated from scratch, incremental revision has been used to repair a pre-computed schedule in response to unanticipated events during schedule execution. The approach analyses the implications of specific schedule features and matches them to behavioral characteristics of appropriate reactive actions that are selected. Our approach uses schedule revision as a unified methodology for both predictive and reactive scheduling and in contrast to most of the above approaches where experimental comparison with other methods has not been reported, our approach has been shown (see section 5) to outperform a variety of other techniques along multiple evaluation criteria.

Our approach integrates Case-based Reasoning (CBR) [10] with constraint-based scheduling. CBR is used for acquisition and flexible reuse of scheduling preferences by storing a user's decisions and schedule revision results as a case. The case is later retrieved when a similar problem is encountered and used to guide selection of appropriate schedule revision, actions (also called repair tactics). Constraint-based scheduling is used to (1) construct an initial (suboptimal) schedule that is then revised through CBR, and (2) to propagate the consequences of case-based schedule modification decisions. A case describes a particular specific experience, which serves as a knowledge structuring mechanism to capture factors that are relevant to the local context for incremental repair and is re-usable in future similar situations. On the other hand, constraint based techniques are necessary to determine the "ripple effects" that spread conflicts to other parts of the schedule as case-based repair actions are applied and specific schedule revisions are made. The integrated methodology for repair-based predictive and reactive scheduling has been implemented in the CABINS system.

The main contribution of this work is the development of a uniform methodology for predictive and reactive schedule optimization according to context-dependent user scheduling preferences. The approach has been validated by experimental comparison with Operations Research and knowledge-based scheduling methods (section 5). The results show that CABINS outperformed the other approaches along a variety of experimental criteria in both predictive and reactive scheduling. The power of the methodology stems from CABINS's ability to interactively acquire cases and use them to generate context-dependent preferences that dynamically guide scheduling decisions.

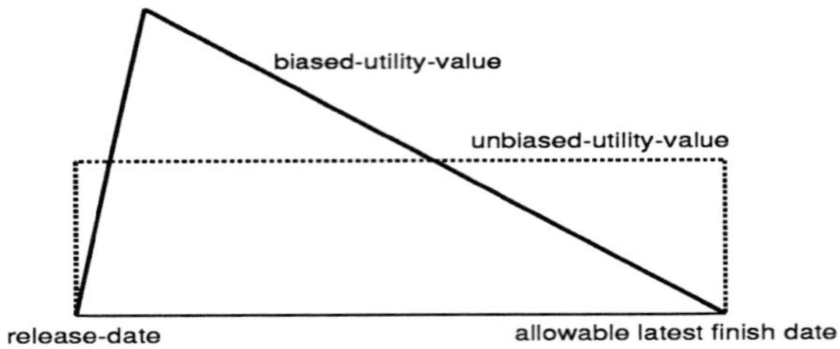

Figure 1. Utility-Functions

2. Constraint-Based Scheduling

The dominant constraints in job shop scheduling are *temporal activity precedence* and *resource capacity* constraints. The activity precedence constraints together with the job release dates and due dates restrict the set of acceptable start times of each activity. The capacity constraints restrict the number of activities that can be allocated on a resource at any moment in time creating possible conflicts among activities competing for the possession of the same resource at overlapping time intervals. The constraint-based component of CABINS views each activity as a *variable.* A variable's *value* corresponds to an activity reservation, i.e. a start time and set of resources needed by the activity. To reduce search, CABINS uses the variable and value ordering heuristics reported in [21]. Variable ordering is done according to the Activity Resource Reliance (ARR) heuristic that checks for the resource/time interval pair that is the most contended for, and the activity that relies most on the possession of that resource over that time interval. This activity is selected as the most critical one and hence is the next to be instantiated (scheduled). The two value ordering heuristics that are relevant to this paper are: (1) Least Constraining Value Ordering (LCV), which selects the reservation that is the *least likely to prevent other activities from being scheduled* according to an unbiased utility function (see figure 1) that reflects no preference for a particular start-time for an activity, and (2) Greedy Value ordering (GV) which selects a reservation based on local start-time and resource preference expressed as a statically predetermined biased utility function associated with each activity (see figure 1). Experiments in [21] using backtracking search have shown that ARR with LCV minimized search but resulted in poor quality schedules, whereas ARR with GV produced near optimal schedules. Backtrack search scheduling with ARR and GV is used to generate schedules for experimental comparison with CABINS (see section 5); ARR with LCV with due date relaxation (when necessary) was used to generate an initial suboptimal schedule to be repaired by CABINS in each of the experimental problems.

3. Reasoning about Repairs

Our approach to incremental schedule modification is influenced by general characteristics common to any computational repair task. Except for simplistic situations, it is generally not possible to bound the scope of a repair in advance. In scheduling, given the tightly coupled nature of scheduling decisions, a change in one part of a schedule often has ripple effects that can span the time horizon of the rest of the schedule from the point of change. This has two consequences. First, each repair action should be kept as localized as possible. Second, the repair process is opportunistic in nature focusing on places in the schedule where constraint violations occur as a result of a prior repair action. The overall objective of the repair task is to balance concerns of schedule quality, stability (salvaging as much of the old schedule as possible at each iteration), and efficiency.

The incremental revision process in CABINS has the following basic steps:

1. Jobs that are suboptimal with respect to the optimization criteria in the initial schedule are identified, sorted (in decreasing order) according to the degree of suboptimality and are repaired according to this sorting. For example, for minimizing tardiness, the most tardy job is the first to be selected for repair.

2. The job under current repair consideration is called the *focal_job* and the activity under current repair consideration is called the *focal_activity*. Activities in a focal_job are repaired in a forward fashion from the first to the last.

3. A repair tactic is selected for repairing the current focal_activity using CBR (see next section) and is applied to the schedule. Application of a repair tactic has four basic parts: First, identify through constraint propagation the activities that will be involved in constraint violations as a result of moving the focal_activity. Second, adjust the reservations of all activities involved in constraint violations by simple right-shifting or left-shifting so that all conflicts are resolved. Third, change the start time utility functions of the affected activities from uniform to triangular, so as to reflect the consequence of the user's preference (see figure 2). Four, un-schedule the focal_activity and all affected activities and re-schedule them using the constraint-directed scheduler with the changed utility functions, so as to bias scheduling to reflect user's preferences.

4. After a repair has been executed, CBR is used to evaluate the repair outcome in the context of the current case-base.

5. If repair is evaluated a success, CABINS selects the next activity to repair, else (if repair outcome is a failure), CBR is invoked to select the next repair tactic for the current activity. The process terminates when there is no more defective job that can be repaired.

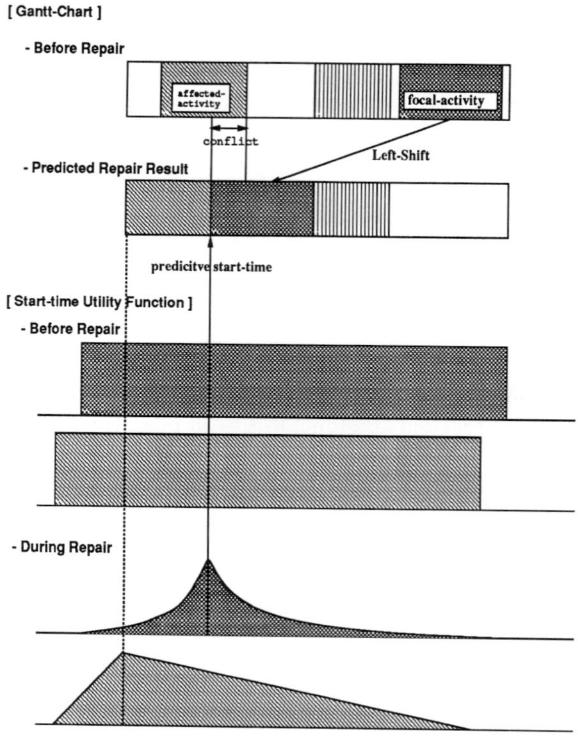

Figure 2. Repair Tactic Application

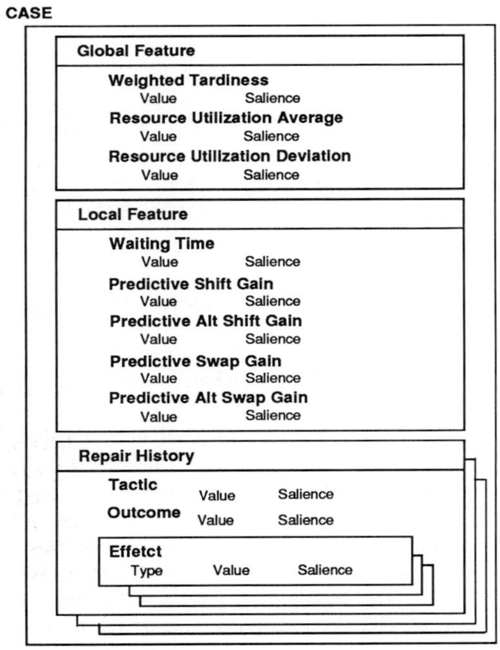

Figure 3. Case Representation

4. Case-based Schedule Repair

Figure 3 shows the information content of a case. Due to the tight scheduling constraint interactions, there is no causal model that can predict which repairs will be effective. The global features give an abstract characterization of potential repair flexibility or the lack thereof for the whole schedule. If, for example, resource utilization is high, it is likely that the schedule is tight, so there is not much repair flexibility. High standard deviation of resource utilization is used as a measure that suggests the existence of bottleneck resources, which in turn makes schedule repairability more difficult. Weighted tardiness is also a global feature since it pertains to the whole schedule.

To place bounds on the repair analysis that guides case-based selection of a local repair action, we define the *repair time horizon*, an interval that temporally spans the focal_activity and is determined as the time interval between the end-time of the activity

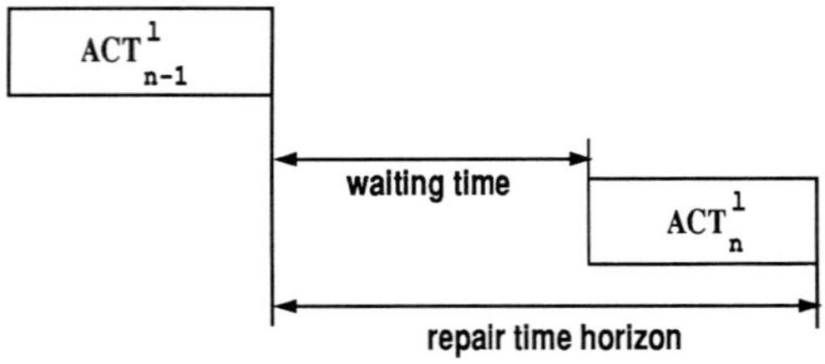

Figure 4. Repair time horizon of focal_activity(ACT_n^l)

preceding the focal_activity in the same focal_job and the end-time of the focal_activity (see figure 4).

Associated with the repair time horizon are local features that are predictive of the effectiveness of applying a particular repair tactic. These features are adapted from those utilized in [17]. For example, predictive-shift-gain (figure 3) is the amount of likely reduction of focal_activity's waiting time when moved to the left within the repair time horizon on the focal_activity's resource. The tactics currently available in CABINS are : *left-shift on the same (or a subtitutable) resource which moves the focal_activity as much to the left as possible within the repair time horizon on the same (or substitutable resource), while minimizing disruption;* swap the focal_activity within the time repair horizon with the activity on the same (or substitutable) resource that causes the least disruption.

CABINS starts with an empty case base. During case acquisition, suboptimal schedules are presented to the user who interacts with CABINS to identify focal_jobs and focal_activities to repair. The user selects out of the available tactics the repair tactic that he deems appropriate to repair a selected focal_activity and applies it to the current schedule. Possible effects are changes in weighted-tardiness of jobs, changes in work-in-process, machine utilization etc. If the user deems the result of a repair tactic application as unacceptable, s/he supplies an explanation in terms of rating the salience of each repair effect, and then attempts application of another repair tactic. The process continues until an acceptable outcome is reached or failure is declared. Failure is declared when there are no more tactics to be applied. The sequence of application of successive repair actions, the effects, the repair outcome and explanation in case of failed application are recorded in the repair history of the case. In this way, a number of cases are automatically accumulated in the case-base. Currently, CABINS has more than 4,000 cases.

4.1. Repair Cycle

The case base is indexed in terms of the global and local features, the repair tactic and outcome and the repair effects. During repair, different sets of these features are used as indices for case retrieval. Once the case base has been acquired, CABINS operates autonomously without further user interaction. After a focal_job and focal_activity have been selected (see previous section), the set of global and local features are used as indices to retrieve past repair episodes to select a repair tactic to be applied. The process of applying a repair tactic is illustrated in section 4.2. After a repair has been applied, repair effects are calculated and evaluation is performed: using the effect type, effect value and effect salience as new indices, CBR is invoked and returns an outcome (acceptable or unacceptable).

If the outcome of current revision is unacceptable, CABINS performs another CBR invocation using as indices the conjunction of the current outcome (unacceptable), the failed heuristic and the case global and local features to find another possibly applicable revision tactic. Invoking CBR with these indices retrieves cases that have failed in the past in a similar manner as the current revision. This use of CBR in the space of failures is a domain-independent method of failure recovery [24], and allows the problem solver to access past solutions to the failure. If the result is acceptable, CABINS proceeds to repair another activity.

To determine case similarity, CABINS uses a k-Nearest Neighbor method (k-NN)[4] where k is a constant (k=5 in the current implementation) according to the formula;

$$Distance_i = \sqrt{\sum_{j=1}^{N}(Salience_j^i \times \frac{CaseFeature_j^i - ProblemFeature_j}{E\_Dev_j})^2}$$

$$Similarity_i = \exp(-Distance_i)$$

where $Salience_j^i$ is the salience of j-th feature of i-th case, (currently salience has been heuristically determined for each feature), $CaseFeature_j^i$ is the value of j-th feature of i-th case, $ProblemFeature_j$ is the value of j-th feature in the current problem, E_Dev_j is the standard deviation of j-th feature value of all cases. In [3] it was demonstrated that such a metric performed well in predicting the outcome in a CBR system with a weak domain theory (foreign trade negotiations). The predictive power of the similarity metric was attributed to the fact that it allows the domain theory to influence judgments of feature relevance in terms of feature salience while retaining the advantages of CBR: allowing partial matches and reasoning in domains where human expertise cannot be formalized.

4.2. An Example

We briefly illustrate the repair process with a very simple example schedule to be repaired shown in figure 5. The example has ten jobs ($J_1, \ldots, J_{10}$) and each job has five activities with the linear precedence constraint. (e.g., O_1^n BEFORE O_2^n, ... , O_4^n BEFORE O_5^n). Resources R_1 and R_2, R_3 and R_5 are substitutable; resource R_4 is a bottleneck. Suppose that the current focal_job is J_8. This job has a weight of 2, a due date of 1250 and the scheduled end-time of its last activity is 1390. Hence it has a weighted tardiness of 2(1390-1250)=280. Suppose the current focal_activity is O_4^8. Since the focal_activity is

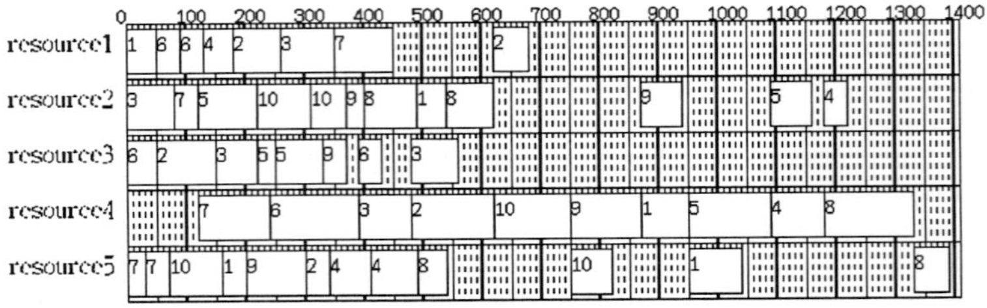

Figure 5. Original Schedule Results

scheduled on the bottleneck machine R_4, which doesn't have any substitutable machine and any idle time in the repair time horizon (time between the end of O_3^8 and the end of O_4^8), CBR invocation selects swap as a repair tactic.

The swap repair tactic calculates the effect on weighted tardiness of swapping the current focal_activity with each activity within the current focal_activity's time horizon and selects the activity that gives the biggest net gain (note that swapping an activity that is scheduled earlier with one that is scheduled later will now delay the earlier activity). In the example, suppose that activity O_4^4 is selected as the activity to be swapped with the current focal_activity O_4^8. Job J_4 has weight 3 and weighted tardiness 3(1370-1320)=150. The effect of applying the swap tactic is that O_4^8 and O_4^4 are unscheduled on R_4 and O_4^8 is re-scheduled to start at time 1090 (the start time of activity O_4^4 prior to the swap). Due to the larger duration of activity O_4^8, now there is the ripple effect of a precedence constraint violation between activity O_4^4 and its successor activity O_5^4 on resource R_2. (In general, many activities could be affected and must be rescheduled as described in section 3). Constraint propagation discovers this constraint conflict and shifts activity O_5^4 further to the right on resource R_2 resulting in the repaired schedule shown in figure 6. The effects of repairing O_4^8 are calculated. CABINS calculates the effects on J_8 and J_4, the jobs affected by the application of the swap on O_4^8. Machine utilization did not change but J_8 had a decrease in weighted-tardiness of 180 time units and a decrease in in-process inventory of 200 units; J_4 had an increase in weighted-tardiness of 150 units and an increase in in-process inventory of 750 units. CBR is invoked using the effect values as indices to determine whether this repair outcome is acceptable.

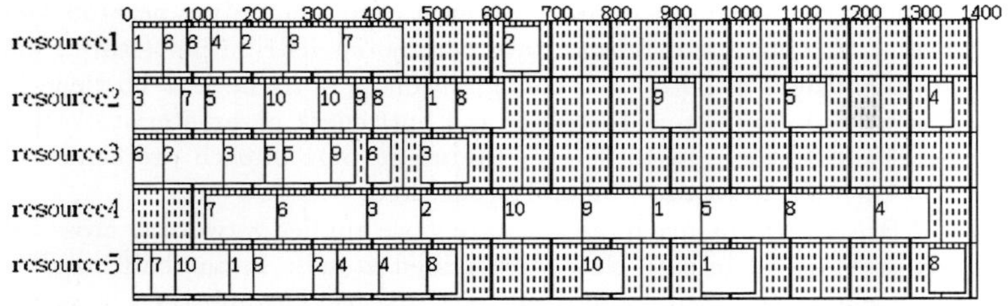

Figure 6. Schedule Results after Repair on O_4^8

5. Evaluation of the Approach

The following criteria have been recognized in the literature (e.g., [17, 25]) as appropriate for evaluating incremental schedule revision: (a) schedule quality with respect to the desired optimization criteria (b) amount of schedule disruption and (c) efficiency of revision. A fourth criterion that has received almost no attention in the literature is incorporation of user preferences in schedule generation and evaluation. These criteria must be balanced and are usually inversely related. In general, either schedule quality is sacrificed to reduce response time and disruption [17], or, when the emphasis is on schedule quality, efficiency concerns are not addressed.

5.1. Predictive Schedule Repair

In predictive schedule repair, the primary objective in our experiments was to optimize schedule quality along different optimization criteria. We compared CABINS with a set of well regarded dispatch heuristics, widely used in manufacturing job shop scheduling, and with a constraint-based scheduler. The dispatch rules selected for the comparison are the Earliest Due Date (EDD) rule, the Weighted Shortest Processing Time (WSPT) rule and the WSPT with job time urgency factor (R&M) rule. These rules have been reported to be particularly good at reducing tardiness under different scheduling conditions. [14]. The constraint-based scheduler (CBS) uses backtrack search, ARR variable ordering, predetermined biased start-time utilities, and GV value ordering (see section 2). In experiments reported in [21], this scheduler achieved near optimal performance with respect to order tardiness and work-in-process-inventory (WIP).

To cover different scheduling conditions, a set of 60 scheduling problems was randomly generated. Each problem has 5 resources and 10 jobs of 5 operations each. Each job has a linear process routing specifying a sequence where each job must visit bottleneck resources after a fixed number of activities, so as to increase resource contention and make the problem tighter. Six groups of 10 problems each were randomly generated using three different values in due date and release date parameter distribution (static, moderate, dynamic), and two values of bottleneck configuration (1 bottleneck, 2 bottlenecks). The slack was adjusted as a function of the range and bottleneck parameters to keep demand for bottleneck resources close to 100% over the major part of each problem. Durations for activities in each job were also randomly generated.

To assess CABINS performance more accurately, we applied a two-way cross-validation method. Each problem set in each class was divided in half. In one half initial schedule was generated using dynamic constraint relaxation (section 2) and then repaired heuristically to gather cases. These cases were used to repair the other half of the problem set. We repeated the above process by interchanging the sample- and the test-set. Our results are the average of the two sets of results. The results in figure 7 show that CABINS outperformed all other methods across both 1 and 2 bottleneck problems in all experiments. The optimization criteria used were weighted tardiness, WIP and the combined objective of minimizing weighted tardiness and WIP.

5.2. Reactive Schedule Repair

One repair strategy to react to unexpected events during schedule execution could be to throw away the rest of the schedule and re-schedule from the point of the occurrence of the unexpected event. It has been speculated (e.g.,[17, 25]) that such strategy may efficiently produce high quality schedules but may increase schedule disruption (though no precise measure of disruption was given in previous work). A second reactive repair strategy could be incremental revision of the existing schedule. The most desirable incremental reactive repair should result in high quality schedules, minimize disruption of factory operation and be *responsive*, i.e. allow continuation of execution as soon as possible. Responsiveness is a very important consideration in factory environments since production delays due to execution failures have a great cost. To date no experimental evidence has been provided to compare these two reactive recovery strategies (incremental repair versus re-scheduling). Our experimental results demonstrate that the incremental repair methodology of CABINS is superior to re-scheduling in response to execution failures along all the desirable evaluation criteria.

Reactive repair in CABINS has the following steps. First, recognize conflicts (temporal and capacity) introduced in the schedule as a result of deleterious unexpected events in the execution environment (e.g. machine breakdowns, delays in operation execution). For example, if an activity finishes later than it was scheduled, there is conflict with the scheduled start time of the next activity on the resource. Second, the effects of the introduced conflicts are propagated downstream (forward in time) from the point in time where the unexpected event happened. For example, if a machine breaks down for some duration d, the activities that were scheduled on that machine for the breakdown interval d must be unscheduled and rescheduled after the end of the breakdown, thus also delaying the activities that were originally scheduled there. The schedule resulting from this step is

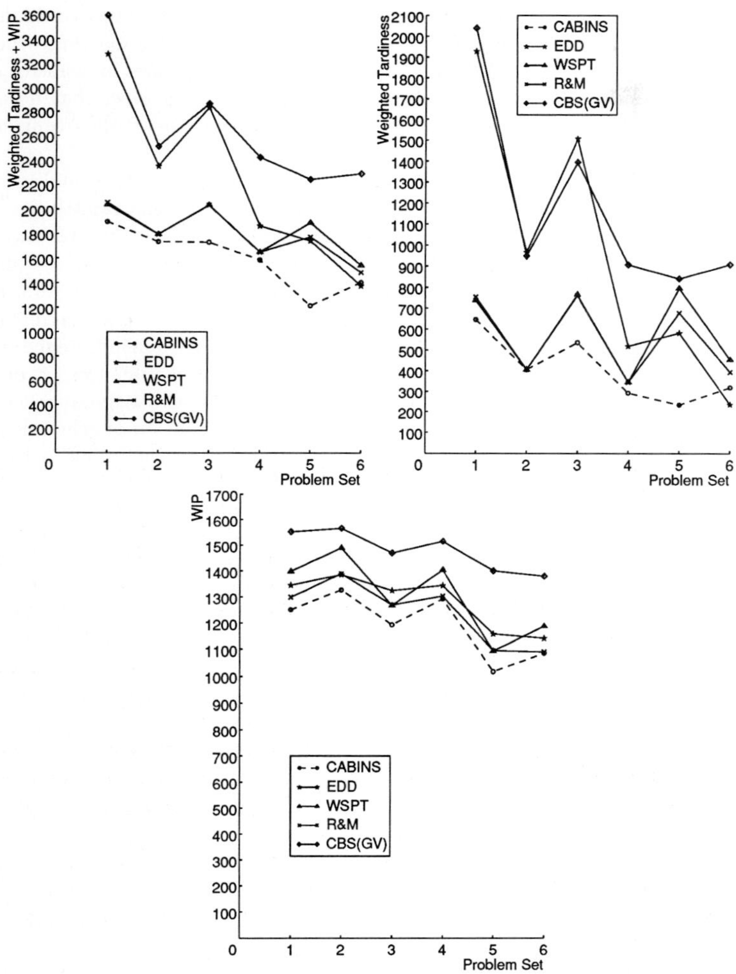

Figure 7. Results of Predictive Schedule

of much worse quality than the predictive schedule before the occurrence of the deleterious unexpected event. In a third step, CABINS is used to repair the suboptimal schedule that resulted in the previous step. The mechanisms that CABINS uses in reactive repair are exactly the same used for predictive optimization (except, of course, that no attempt is made to repair activities that have being already executed prior to the occurrence of the unexpected event).

We evaluated the performance of CABINS with respect to schedule quality, minimization of disruption and responsiveness as compared with the constraint-based scheduler (CBS) with backtrack search, predetermined utilities and GV value ordering. In each experiment, CABINS was used to repair a predictive schedule in response to unexpected loss of resource capacity (machine breakdown). We defined three schedule disruption measures: (1) difference of start times between the repaired and the original predictive schedule (before occurrence of the unexpected capacity loss), (2) difference in resource assignment of activities in the repaired versus the original schedule, and (3) difference in sequencing of activities on a resource in the repaired versus the original schedule. Such changes could cause serious difficulties in the smooth continuation of schedule execution.

We compared the performance of reactive repair versus re-scheduling in two different machine breakdown scenarios, one and two bottleneck problems. Each scenario has 10 sets of problems. In both scenarios, the machine breakdown was on the busiest non-bottleneck machine. Breakdown duration was made 10 times the average activity duration. Minimizing weighted tardiness was selected as the optimization criterion (since the long duration of the breakdown made weighted tardiness the dominant concern). The average results of solved problems [1] in the first set of experiments are shown in the left table in Tables 1 and the average results of solved problems in the second set of experiments are in the right table.

The results show that reactive repair clearly outperforms re-scheduling in minimizing disruption along all three disruption measures and also maximizes schedule quality. On the other hand, eactive repair is much slower than rescheduling. However, this apparent inefficiency is misleading.

The reason is the rapid and monotonic repair behavior of CABINS. Graphs in figure 8 show repair responsiveness of CABINS in 2-bottleneck problems 1, 4 and 7 (where CABINS had the lowest repair efficiency). These graphs show that, even in those three problems, CABINS achieved better quality than CBS *at the time point when CBS finished re-scheduling.* From the left graph in figure 8, for example, we see that CBS finished in 9.3 seconds, At 9.3 seconds CABINS has reduced weighted tardiness to 1430 units compared to 1560 units achieved by CBS. Since, after each repair, CABINS produces an *executable* schedule, if the repair process is stopped after 9.3 secs, the produced schedule is of higher quality than the schedule produced by re-scheduling and, furthermore it can be executed. This is a very important characteristic in manufacturing environments where

[1]Re-scheduling by CBS could not solve all problems, a serious drawback in real factory environments, while CABINS reactive repair is always guaranteed to produce an executable (conflict-free) solution.

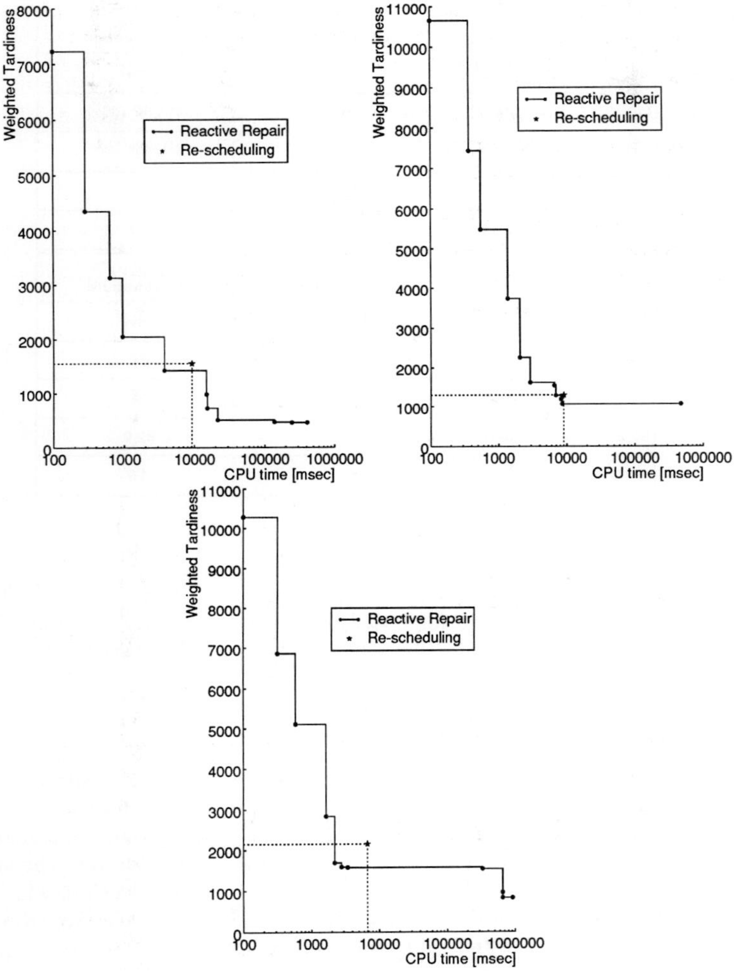

Figure 8. Repair Responsiveness of CABINS in Reactive Situations

		CABINS reactive-repair	CBS(GV) ** re-schedule
Disruption	Start Time	7600	10460
	Routing	10	12
	Sequencing	23	29
Repaired Weighted Tardiness (%)		97.6	91.2
CPU time (second)		268.8	8.6

		CABINS reactive-repair	CBS(GV) ** re-schedule
Disruption	Start Time	5150	7490
	Routing	9	11
	Sequencing	19	24
Repaired Weighted Tardiness (%)		99.9	92.0
CPU time (second)		77.1	4.9

Table 1
Reactive repair vs Re-scheduling

rapid resumption of production, without sacrificing schedule quality, is critical. Therefore, we see that CABINS outperformed CBS re-scheduling on quality, disruption and execution responsiveness simultaneously.

6. Conclusions

In this paper, we advocated a unified framework for knowledge acquisition and iterative revision for predictive and reactive schedule optimization. The approach utilizes CBR-based mechanisms for recording user preferences, repair tactics and explanations, and constraint-based scheduling for application of the selected repair tactics. The approach is predicated on (a) the existence of a set of schedule repair tactics, each of which operates with respect to a particular local view of the problem and offers selective advantages for improving schedule quality, (b) on capturing user scheduling preferences and judgments, and (c) on retrieving and re-using these preferences to dynamically change scheduling utilities during revision. Our experimental results show that the approach (1) was able to capture and effectively utilize user scheduling preferences that were not present in the scheduling model, (2) outperformed other scheduling methods along mulitple evaluation criteria in both predictive and reactive contexts.

REFERENCES

1. E. Biefeld and L. Cooper. Bottleneck identification using process chronologies. In *Proceedings of the 12th International Joint Conference on Artificial Intelligence (IJCAI-91)*, Sydney, Australia, 1991.
2. J. R. Bitner and E. M. Reigold. Backtrack programming techniques. *Communications of the ACM*, 18(11):651–655, 1975.
3. Pazzani M. Cain, T. and G. Silverstein. Using domain knowledge to influence similarity judgements. In *Proceedings of the 1991 Case-Based Reasoning Workshop*, pages 191–199, Washington, D.C., 1991.
4. Belur V. Dasarathy, editor. *Nearest Neighbor (NN) Norms: NN Pattern Classification Techniques.* IEEE Computer Society Press, Los Alamitos, CA, 1990.
5. Rina Dechter and Judea Pearl. Network-based heuristics for constraint satisfaction problems. *Artificial Intelligence*, 34(1):1–38, 1988.
6. Mark Fox. *Constraint-Directed Search: A Case Study in Jop Shop Scheduling.* PhD thesis, Department of Computer Science, Carnegie Mellon University, 1983.
7. Michael R. Garey and David S. Johnson. *COMPUTERS AND INTRACTABILITY: A Guide to the Theory of NP-Completeness.* W. H. FREEMAN AND COMPANY, New York, NY, 1979.
8. Robert M. Haralick and Gordon L. Elliott. Increasing tree search efficiency for constraint satisfaction problems. *Artificial Intelligence*, 14(3):263–313, 1980.
9. Naiping Keng and David Y. Y. Yun. A planning/scheduling methodology for the constrained resource problem. In *Proceeding of the Eleventh International Joint Conference on Aritificial Intelligence*, pages 998–1003, Detroit, MI, 1989. IJCAI.
10. J. Kolodner, R. Simpson, and K. Sycara. A process of case-based reasoning in problem solving. In *Proceeding of the Ninth International Joint Conference on Aritificial Intelligence*, pages 284–290, Los Angeles, CA, 1985. IJCAI.
11. A. K. Mackworth. Consistency in networks of relations. *Artificial Intelligence*, 8(1):98–118, 1977.
12. K. Mckay, J. Buzacott, and F. Safayeni. The scheduler's knowledge of uncertainty: The missing link. In *Proceedings of IFIP Working Conference on Knowledge Based Production Management Systems*, Galway, Ireland, 1988.
13. S. Minton, M. D. Johnston, A. B. Philips, and P. Laird. Solving large-scale constraint satisfaction and scheduling problems using a heuristic repair method. In *Proceedings, Eighth National Conference on Artificial Intelligence*, pages 17–24, Boston, MA., 1990. AAAI.
14. T. E. Morton, R. M. Rachamadugu, and A. Vepsaleinen. Accurate myopic heurisrics for tardiness scheduling. Technical Report 36083-84, Graduate School of Industrial Administration, Carnegie Mellon University, 1984.
15. N. Muscettola. Hsts: Integrating temporal planning and resource allocation scheduling. In M. Fox and M. Zweben, editors, *Knowledge-Based Scheduling.* Morgan Kauf-

mann, San Mateo, CA, 1992. Forthcoming.

16. N. Muscettola. Scheduling by iterative partition of bottleneck conflicts. In *Proceedings of the Ninth Conference on AI Applications*, Orlando, Fla., March 1993. IEEE.
17. P. S. Ow, S. F. Smith, and A. Thiriez. Reactive plan revision. In *Proceedings of the Seventh National Conference on Artificial Intelligence*, pages 77–82, St-Paul, Minnesota, 1988. AAAI.
18. Claude Le Pape and Stephen F. Smith. Management of temporal constraints for factory scheduling. Technical Report CMU-RI-TR-87-13, The Robotics Institute, Carnegie Mellon University, 1987.
19. Paul W. Purdom, Jr. Search rearrangement backtracking and polynomial average time. *Artificial Intelligence*, 21:117–133, 1983.
20. Patrick Prosser. A reactive scheduling agent. In *Proceeding of the Eleventh International Joint Conference on Aritificial Intelligence*, pages 1004–1009, Detroit, MI, 1989. IJCAI.
21. Norman Sadeh and Mark S. Fox. Variable and value ordering heuristics for activity-based job-shop. In *Proceedings of the Fourth International Conference on Expert Systems in Production and Operations Management*, pages 134–144, Hilton Head Island, SC, 1990.
22. Stephen F. Smith. Knowledge-based production management: Approaches, results and prospects. Technical Report CMU-RI-TR-91-21, The Robotics Institute, Carnegie Mellon University, 1991.
23. Stephen F. Smith, Peng Si Ow, Nicola Muscettola, Jean-Yves Potvin, and Dirk C. Matthys. An integrated framework for generating and revising factory schedules. *Journal of the Operational Research Society*, 1990.
24. K. Sycara. *Resolving Adversarial Conflicts: An Approach Integrating Case-Based and Analytic Methods.* PhD thesis, School of Information and Computer Science Georgia Institute of Technology, Atlanta, GA, 1987.
25. M. Zweben, M. Deale, and M. Gargan. Anytime rescheduling. In *Proceedings of the DARPA Workshop on Innovative Approaches to Planning, Scheduling and Control*, pages 251–259, San Diego, CA., 1990. DARPA.

Knowledge-Based Reactive Scheduling (B-15)
E. Szelke and R.M. Kerr (Editors)
Elsevier Science B.V. (North-Holland)

Reactive Scheduling—An Intelligent Supervisor Function

E. Szelke* and G. Márkus**

*Computer & Automation Institute, Hungarian Academy of Sciences,
P.O. Box 63, Budapest 112, H-1518

**AblakSoft Ltd., Horánszky u. 26., Budapest, H-1085

Abstract

In dynamic manufacturing environments, no matter how well defined a predictive schedule is, reactive scheduling is needed to complement it by managing changes due to unexpected events/disturbances on the shop floor. A supervisory control system is being developed with an intelligent supervisor unit to provide various intelligent assistance capabilities relying on a cognitive model of operator functions. In the system, reactive scheduling is conceived as an intelligent supervisor function. It assists the human by revising an active but flawed schedule and adjusting it to changed system conditions. An underlying blackboard control architecture provides the proper frame for problem solving by the intelligent supervisor unit and for the cooperation of case and rule-based reasoning components, to select adequate repair policy, and incrementally resolve conflicts in the schedule currently under revision with attendance to the whole system performance.

Keyword Codes: H.1.2; I.2.4; I.2.6; I.2.8
Keywords: User/Machine Systems; Knowledge Representation Formalisms and Methods; Learning; Problem Solving, Control Methods, and Search

INTRODUCTION

In live industrial environment involving dynamic manufacturing processes, effective shop-floor control and scheduling can make a major contribution to the competitive power of a company to meet customer demands as well as utilize its assets/resources. As stated in Khaw *et al.* (1991), "companies in the world would come to the conclusion that the battleground where competitive wars between companies are being fought is the shop floor." Unanticipated events on the shop floor such as machine breakdown, expedition of critical new orders, supplier delivery changes (resulting in lack of materials), product quality assurance and rework, all form parts of the environment's dynamic and stochastic nature.

Dynamic scheduling approaches represent a first attempt to overcome the complexity of industrial shop floor scheduling problems and make scheduling results more applicable in practice. These approaches pertain to the determination of real-time production schedules generated as near to the time of schedule execution in manufacturing processes as possible (Sarin and Salgame 1989, Szelke and Mészáros 1990). To produce more robust solutions to (predictive) scheduling problems in dynamic decision making environments, some systems use the principle called "least commitment strategy" (Kerr 1991).

However, even the best quality predictive schedules may become obsolete by the time of their execution due to many unpredictable changes in the factory floor. The highly time varying preconditions in the production environment mentioned above need continuous maintenance of the predicted schedules to keep them in line with unfolding events in the shop floor (Kerr 1991). For this reason, a reactive scheduling function is inevitably needed in real-time shop floor control systems which have on-line feedback from the executed manufacturing processes to reflect the actual manufacturing situations.

To cope with the above shop floor contingencies, it is likely that an effective *integration and coordination* of shop floor *decision making and control functions* are necessary, *with* the *implication of human factors*, to provide a required level of the control system's "intelligence" in response to rapid changes in the manufacturing environment.

With respect to the requirements above, a supervisory control system with various intelligent assistance capabilities, among them a knowledge-based reactive scheduling ability, is proposed as a viable alternative. In the paper the term *supervisory control system* denotes a control system that integrates the human operator, by design, in the control system which performs: detailed scheduling of shop floor activities and real-time control during execution of the planned activities, with a *reactive scheduling ability* to prevent or compensate for disturbing effects of unanticipated events during the manufacturing process.

The required *underlying integrating framework* for shop floor decision making and control is provided in the system by the implementation of different paradigms, a major example being the *supervisory control paradigm* (Sheridan 1982, Mitchell 1988). This explicitly addresses the role of human, computerized algorithms, and knowledge-based system components, and integrates the three factors, by design, to enhance the whole system's performance against the actual target of production.

While implementing the supervisory control paradigm, the system design strives to exploit the strengths of both the human and computer components of the complex system to achieve some degree of system intelligence. Here *intelligence* means the ability of *understanding* (i.e., of transferring information into properly coded knowledge) and the ability of *inferring* (i.e., of transferring acquired knowledge into possible decisions or choices). Both abilities are enabled in the computerized system by a cognitive model of operator functions, complemented with intent inferencing properties and embedded in a blackboard control architecture for implementation.

Although some decision making and control tasks/functions can be completely automated, many control functions in complex manufacturing environments require a design that incorporates human override capabilities. The latter is also considered in the proposed system at different phases of decision making.

Within the system, an *intelligent supervisor unit* is being developed for providing the human supervisor with on-line assistance in shop floor control and decision making processes. By an intelligent interface design to the system, it is ensured that the human operator remains an integral part of the system and is able to perform effectively when needed. The knowledge based intelligent supervisor unit, from an operator's assistant perspective, can interact dynamically with a human operator/supervisor. The cognitive model of operator functions, which has been gradually collected from industrial practice, is being developed and utilized in the system's intelligent interface to the user (human operator). Via this normative model of operator functions, the user (operator) interface incorporates the semantics of the control domain and the supervisory tasks, thus, through its design, it incorporates some degree of intelligence. It is being built to assist the human while he/she performs or instantiates different planning/scheduling, monitoring, intervening or administering functions during the automated process execution.

Reactive scheduling is an *intelligent supervisor function* in the above context, for revising active but flawed schedules by adjusting them to changed system conditions. This function utilizes knowledge-based (combined case and rule-based) opportunistic reasoning for schedule repair applying an appropriate blackboard model of problem solving. Opportunistic or hypothetical reasoning is an inherent feature of the blackboard models of problem solving (Nii 1986), and thus also characterizes the proposed system's solution approach to reactive scheduling.

For giving more *flexibility in modeling and problem solving* and to eliminate the inefficiency of simple rule-based knowledge systems that lack learning capability, a case-based learning component (and module) is involved in the system to ensure a continuous improvement of the knowledge-base/case-base. The related option of the interface to system specialist (ISS) enables the system to evaluate/learn successful schedule revision policies by examples and to accumulate knowledge associated with handling new events and related decision procedures. The ISS option allows also the maintenance/modification of the controlled system's model and the knowledge-base/case-base, thus ensuring the system's broader applicability, with a likely open system feature. This is of particular importance in computer integrated manufacturing environments today (Rasmussen 1990), where tasks and functions of people are not stable, and introduction of advanced tools into manufacturing processes can influence work procedures at the shop floor, roles of people, and organizational structures.

The paper is organized as follows. In Section 1, some major features and concepts of reactive scheduling, as a relatively new principle, are described. In Section 2, the functional structure and the main components of the supervisory control system with their interrelations which generate reactive scheduling decisions are given. Then in Section 3, the underlying knowledge representation and modeling approaches to problem solving in reactive scheduling within the proposed system are discussed. Section 4 is focused on the utilized blackboard problem solving methodology. Finally, the Conclusions section refers to some open problems that are subjects of future research.

1. MAIN FEATURES AND CONCEPTS OF REACTIVE SCHEDULING

With respect to dynamic manufacturing environments, the *scheduling problem's complexity* lies in the requirements to assign jobs to machines on the shop floor for *meeting* a set of often *conflicting objectives* while *not violating* a set of *precedence, time and resource constraints*. Scheduling problems have generally a combinatorially explosive search space at the outset (they are computationally NP-hard), which can usually be reduced to a search space of feasible solutions by application of constraints. Besides the combinatorial complexity, the time varying nature of objectives and that of the resources' availability, etc. result in additional complexity of the scheduling problem in live manufacturing environments.

For the above complex problems, the existing *conventional approaches* to general job-shop like problems, those based on different Operations Research (OR) techniques are insufficient. Due to the many simplifying assumptions used in their models, they are difficult to customize to company-specific variety of local manufacturing situations. They are not applicable in industrial practice because of prohibitive computing times and their static, deterministic approach to scheduling. Other heuristic techniques using priority dispatching rules for assigning jobs to machines can generally result in only local optima and not achieve good global performance. Limitation of conventional OR-based scheduling procedures lies in their inability to address either the multifaceted goal structure of manufacturing processes or the transient and dynamic nature of real-time scheduling and control problems. Moreover, analytic OR techniques are incapable of representing some of the qualitative, though very significant, features of the control and scheduling problem. For example, conflicting objectives (Berry 1990) of maximizing system throughput and minimizing WIP are poorly addressed by procedures which seek to optimize a single performance measure. Also, analytic methods may require such computation time as to produce results which are of little or no use at the time of availability.

Conventional approaches utilize in their scheduling models only a *fragment of the scheduling knowledge* which exists in practice on the shop floor. Therefore, their models lack the level of fidelity and also the adaptability to changes, i.e., flexibility. As pointed out by Fox (1983), practical factory scheduling problems are affected by a large number of domain specific constraints most of which are not readily expressible in conventional OR-oriented forms of representation.

AI approaches represent an important step toward the required expressiveness, flexibility and responsiveness in handling realistic scheduling problems. These approaches provide rich knowledge representational forms for describing, propagating and, if necessary, relaxing the many constraints that apply to schedules in live environments. They allow a variety of problem decomposition and search strategies to assist in finding a satisfactory solution.

AI-based scheduling approaches generally construct a search tree to help find a feasible solution to the problem. The search tree then explored in a depth-first or a breadth-first mode depending on the application (see, e.g., Atabakhsh 1991, Kerr 1991). Various methods, among them the constraint directed search techniques (Ow *et al.* 1988), can be applied to reduce the size of the tree and to achieve a feasible solution with the lowest effort.

Knowledge-based systems represent advances in incorporating several kinds of knowledge available in different forms but unused by conventional approaches (e.g., Bensana *et al.* 1988). However, crucial problems are likely to be the appropriate representation of constraints and uncertainty.

Reactive scheduling has emerged as a relatively *new concept* in production planning and control, attracting the interest of both academic and industrial researchers to utilize available knowledge-based techniques to the problem or provide advanced tools for subsequent industrial applications. The problem lends itself to the application of advanced knowledge-based techniques including learning as well.

The *problem of reactive scheduling* is concerned with repeated and quick contingency-driven reactive decision making for handling unexpected events/uncertainties in dynamic manufacturing environments. It generally supposes an existing predictive schedule, good enough to be relatively robust against unexpected events on the factory floor. Reactive scheduling aims at re-using the undisturbed parts of the original schedule and adjusting only those parts that are affected by disturbances, whilst minimizing disruptions to the executed process. Therefore, the problem is highly complex due to the mostly uncertain information/knowledge available on the system status, valid goals, etc., and cannot be captured by any kind of analytic methods. Only knowledge-based approaches that lead to satisfying solutions can be taken into consideration, and these also require an intelligent utilization of human knowledge/expertise in addition to the propositions by the knowledge-based system (Sycara and Miyashita 1993, Szelke and Márkus 1993).

Knowledge about facts (status of system's resources and jobs/orders) *and domain knowledge* with rules/frames/nets that are invoked to make decisions as a result of changes in system status, need to be readily available in the knowledge-base. To utilize this knowledge in the course of problem solving and to control the reasoning process, some inferencing mechanisms and case indexing/retrieval techniques are also required.

Knowledge representation and effective domain modelling are vital to the design of intelligent reactive scheduling systems operating in real-time. This needs the development of proper data structures so that the knowledge can be conveniently stored, tested, retrieved and updated for problem solving with minimal time consumption. Knowledge representation techniques utilized in reactive scheduling vary from simple production rule representations (Bruno *et al.* 1986), through frame and schema representations (Smith 1987) to hybrid representations generally implemented by means of Object-Oriented Programming (Smith 1993).

A recent overview of the knowledge-based reactive scheduling domain's literature (Szelke and Kerr 1993) has revealed the application of diverse *AI-based reasoning approaches to the problem* such as:

- ❖ macro and micro opportunistic reasoning (ISIS, OPIS, BOSS and MICRO BOSS systems referred to by Fox 1983, Ow *et al.* 1988, Hynynen 1989, Sadeh and Fox 1990),
- ❖ incremental constraint satisfaction, described by Elleby *et al.* 1988,
- ❖ opportunistic scheduling with real-time blackboard-based architecture referred to by Lalanda *et al.* (1992),
- ❖ approaches using genetic algorithms (Lawton 1992),
- ❖ neural-network-based approaches (Bienner 1992, Hoong *et al.* 1991, Khaw *et al.* 1991),
- ❖ case-based reasoning combined with other approaches (Sycara *et al.* 1993, Szelke *et al.* 1993),
- ❖ combined approaches using simulation and learning (Kerr and Kibira 1993),
- ❖ fuzzy set theory combined with other approaches (Yuan and Wu 1991, Kerr and Walker 1989).

With respect to the *requirements on knowledge-based reactive scheduling systems*, some peculiar features can be extracted from the reviewed literature (Kerr 1991, Noronha and Sharma 1991, Kempf *et al.* 1991, Laffey *et al.* 1988, Atabashkh *et al.* 1991), and also some features vital for supervisory control environments (Sheridan *et al.* 1982, Mitchell 1988). They are collected below:

- ❖ Schedules should be maintained in line with unfolding shop floor events.
- ❖ A real-time response capability is required including the option of selecting a viable response to the situation predicted to exist by the time the response has been computed and enacted.
- ❖ The original schedule should be modified to the minimum extent possible.
- ❖ Reactive scheduling should be able to handle uncertain knowledge and incomplete information.
- ❖ Different kinds of knowledge, including a wide variety of local scheduling knowledge possessed by a skilled shop floor manager, and also generic knowledge of scheduling theory and OR/AI techniques possessed by AI/OR experts should be included in decision making.
- ❖ The system should manifest flexibility in knowledge representation, which must be selected with respect to its usability in real-time decision making environment.
- ❖ Predictive and reactive scheduling are complementary functions in industrial shop floor control systems.

The requirements above should be complemented by a more comprehensive view of a reactive scheduling system design for real-time and interactive systems, where the appropriate design of the user interface is a crucial problem influencing the whole knowledge-based system performance. The conception of distributed cooperating intelligent agents in reactive scheduling related decision making demands some further considerations detailed in Smith (1993).

The above mentioned state-of-the-art overview (Szelke and Kerr 1993) on the domain of knowledge-based reactive scheduling has indicated two contrasting approaches to the problem. One approach considers reactive scheduling as a problem closely coupled with predictive scheduling, in an integrating framework. The other approach conceptualizes reactive scheduling as a real-time problem emerging in the execution of shop floor processes, and hence it stresses response time considerations. Interactive and/or distributed multi-agent environments give additional complexity to the problem.

The vast majority of control systems in which the implementation of reactive scheduling might be anticipated are real-time only in the simplest sense: the programs run fast enough to keep up with the plant's events. The systems are event sensitive. Real-time computation must include the option of selecting a response to the situation that will exist by the time the response has been computed and enacted. The computation time for the response can be reduced by storing a large number of predefined responses, but this is at the expense of storage space. The implementors of knowledge-based reactive scheduling systems for real-time applications must make decisions about trade-offs between computation time and storage space (Laffey *et al.* 1988). Effective reactive scheduling generally implies low time and high space. The time required to construct sound responses might be further reduced by stoi ing some search control knowledge, demanding space.

Case-based reasoning might be regarded as particularly relevant here (Kolonder *et al.* 1985), since part of the rationale of this approach is to retrieve previously stored responses and adapt them as necessary. Viewed in terms of the above perspective, case-based reasoning requires medium time and space.

Within the proposed joint man-machine cognitive system (Woods 1990) for supervisory control of dynamic manufacturing processes, the reactive scheduling problem is considered rather as a real-time problem to be solved by interactive cooperation of the automated intelligent assistant called *intelligent supervisor unit* (ISU) of the system (with a blackboard model and distributed knowledge-sources for case and rule-based reasoning), and the human operator who is integrated into the control process. The system described in the following section, is designed to perform in essence an opportunistic scheduling activity, within the blackboard control architecture selected for problem solving.

2. FUNCTIONAL STRUCTURE OF THE SUPERVISORY CONTROL SYSTEM

Efficient utilization of information and knowledge within the pi oposed supervisory control system is attained by a reasonable *implementation of different important paradigms*, detailed in (Szelke 1990) and briefly outlined below, such as:

- adaptivity of the whole control cycle of the system
- a blackboard-based opportunistic reasoning approach to reactive scheduling
- the supervisory control paradigm providing an integrating framework for control and decision making in the system
- the CIM philosophy to integrate (in the future) different distributed decision agents of shop floor processes (concerned with supervising auxiliary processes/activities, such as preparatory engineering, preventive maintenance, etc.)

❖ intelligent interface design guidelines/propositions.

The *functional structure of the supervisory control system* with its main sub-systems, modules, knowledge-bases, etc. are given by indicating their interrelations within the real-time shop floor control layer, and illustrated by Figure 1.

From the aspect of reactive scheduling, the following modules/sub-systems contribute to the control and decision making processes within the adaptive control loop which contributes to state identification, decision making and control as follows:

❖ *Identification* of system status and the corresponding decision situation are carried out by shop floor data/event acquisition (based on on-line monitoring and diagnosis of process control operating at equipment/sensor level) and a comparison-evaluation on predicted and performed actions. This evaluation may reveal divergencies from plans and/or inefficiency in performance. It also implies *analysis* to detect new conflicts that may have just arisen (if any). The results are offered for the human operator to check for the type of disturbance and to reveal its effect on all downstream operations by activating a preview simulation run. A possible decision situation (as a result of this analysis) can be characterized by:
  - ❑ the status of all resources concerned with processing, such as machines, operators, transport devices, buffers, load/unload stations, tools-palettes-fixtures, etc.,
  - ❑ the degree of completion of the previously scheduled orders,
  - ❑ the degree of implementation of the current schedule,
  - ❑ the current set of conflicts,
  - ❑ the type of current disturbance,
  - ❑ the effect of the current disturbance on resources/constraints/objectives,
  - ❑ the current set of goals,
  - ❑ the current combination of newly arrived customer orders/order types and due dates,
  - ❑ etc.

  As new orders arrive, the schedule will have to be revised, based on the new order type and due date, and will also have to be progressively "rolled" down the time horizon by preview simulation.

❖ *Decision making* to ensure the continuity of the executed process is based on the results coming from the above detailed comparison–evaluation–analysis. The system itself might propose modification of control policy (e.g., to change performance criteria and corresponding rules selection) which might also be checked and overridden by the human operator.

  With respect to the indicated modifications, the intelligent supervisor unit yields a set of hopefully appropriate repair strategies from the system's case-base/knowledge-base (using preference ranking) and initiates a reactive schedule session for each strategy selected, resulting in a list of alternative solutions for the human operator to compare and choose from.

❖ *Control*, for execution in the process control layer, is accomplished by the human operator giving commands or by the intelligent supervisor unit (ISU) itself in case of timeout.

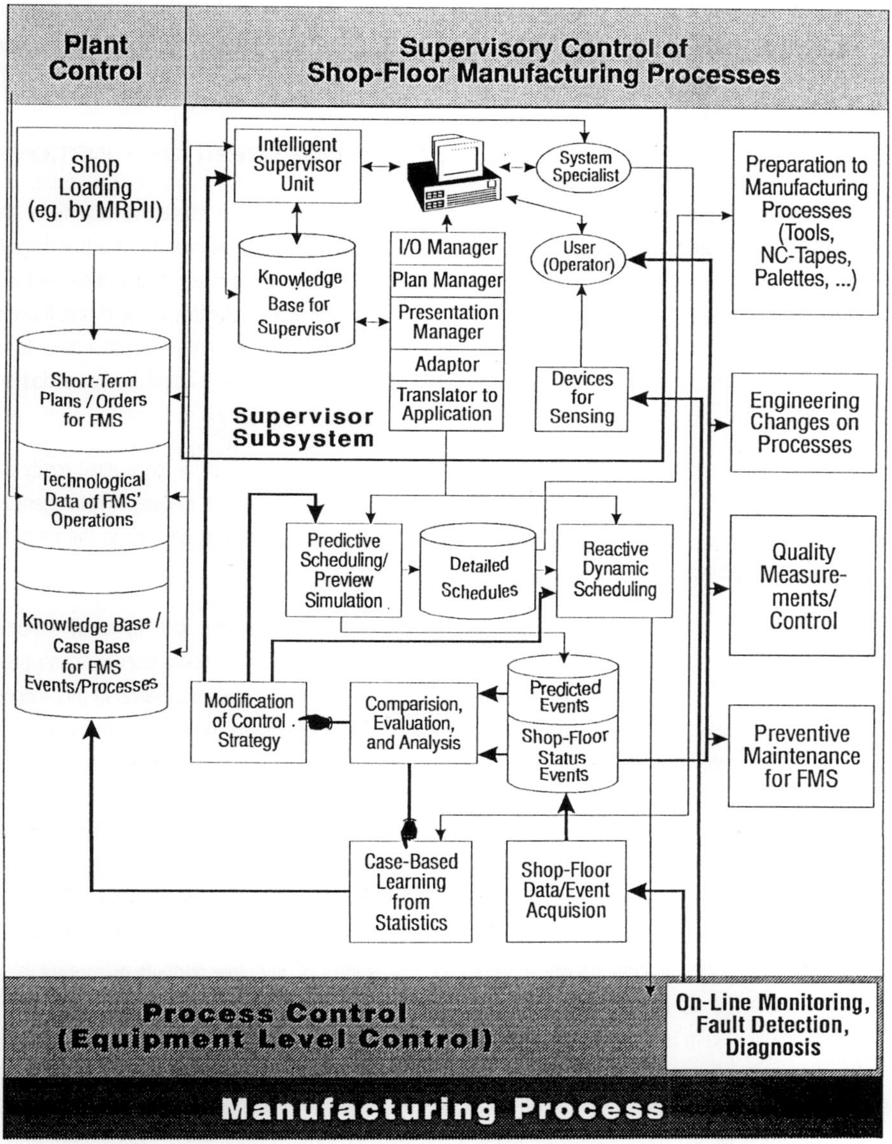

Figure 1 Scheme of the Supervisory Control System's Structure

In addition to identification, decision making and control, the system's *learning* module allows the handling of novel situations with the help of the human operator and with subsequent inclusion into the case-base. Clustering of these unclassified cases is left to the system specialist.

3. KNOWLEDGE REPRESENTATION AND MODEL OF OPERATOR FUNCTIONS

For selecting the proper scheme of knowledge representation for the real-time interactive system, we start from demands for providing the human with *sufficient and timely information* pertaining to the actual decision situation. With respect to the above and general flexibility requirements (for, e.g., easy modification/extension as the available knowledge changes) a frame/object representation was found to be the proper choice from the outset. Primarily, a cognitive model of operator functions was established and, corresponding to its abstraction hierarchy, a proper blackboard architecture built in which the model could be embedded and implemented.

Knowledge engineering considerations indicated that the blackboard architecture lends itself easily to a frame/object/schema type of knowledge representation, which can also best match the use of heterogeneous knowledge sources (e.g., combined use of rule-based and case-based knowledge sources in the proposed system) allowed by blackboard models.

The ISU of the system should use procedures/representations compatible with those of a human scheduler. It is observed that a human scheduler typically groups the plant knowledge based on concepts or objects. *Knowledge frames* provide a mechanism to represent different pieces of information in this manner. Thus, the knowledge representation schemes used in the system include frame (object) structures and some relational tables. Some *major frame structures* have been taken into consideration—the ones describing the controlled system model and the manufacturing tasks are the following: (1) *job/order frames* for organizing all knowledge attributes associated with a job (characteristics of jobs, those of manufacturing operations on a job, statuses of different jobs such as processed/waiting/finished, orders remaining time until due date, route alternatives, etc.); (2) *plant/shift information frames* (on a shift basis, at daily start, three frames are to be created, one for each shift, within the production time window/period of predictive scheduling, usually one day, the current). *Temporal attributes* whose values are independent of any other attribute or concept except time are considered as global temporal attributes. Among the global temporal attributes, there is a relationship. The system is designed for quasi real-time control in the sense that the temporal attributes are updated intermittently on the blackboard whenever a dynamic situation (an event) is encountered and the system's status has just changed. Data which are relational in nature are stored in a structure called a relations table, which represents a particular object and is a collection of attribute-value tuples. The objects may be physical entities such as machines, or they may be conceptual entities such as "machine availability". Attributes are associated with only one particular characteristic or property of the object.

Effective automated assistance in decisions *requires* a thorough *understanding of the human role* in such systems. The automated nature of the control system makes most operator activities primarily cognitive rather than physical (Woods 1988). Thus, models of operator decision making must represent cognitive as well as physical activities. Such models can also provide intelligence in information display design for supervisory applications (Mitchell 1986, Rouse 1987), which is also considered by the proposed system.

The cognitive model of operator functions is in essence a discrete control modelling construct to describe and normatively prescribe operator behavior in a range of supervisor activities. The cognitive model utilized in the system represents the interrelations between dynamic system states and operator functions. For given states, operator functions are represented hierarchically, and individual functions are decomposed into subfunctions, tasks, control actions and information requests.

The model is built as *a heterarchic-hiearchic network* of nodes to represent operator control functions (see Figure 2). The network heterarchy represents the collection of high level functions that comprise the operator's role in system control. The hierarchy defines the decomposition of high level functions into subfunctions, tasks, and actions that the operator uses to coordinate both control actions and information selection and integration. The dynamic nature of the system is defined by the arcs between network nodes. Each node is represented as a finite-state automaton. Thus each node has a state space, an input alphabet and an output alphabet. The network structure constrains the inputs. Inputs to a node are outputs from higher level nodes, or exogenous events in the case of the top network level; outputs define the outputs of the operator function model. Outputs can take either of the two forms: manual control actions or cognitive information seeking and decision making actions. At each node, a state space and a next state transition function determine how a new input causes the state to change, given the current system status. State transition functions for the model are typically nondeterministic in form (i.e., for a given system state and current input, the next state transition function maps the current state into a set of possible next states). In this way, the model accounts for both the flexibility in many operator control decisions (e.g., in reactive scheduling decisions by alternatives) and the constraints inherent in particular system configurations.

The skilled operator of a complex system has a great deal of knowledge that is used to carry out operator functions. An operator's knowledge has several forms such as *static knowledge* about system components and connections between them (e.g., feasible structural system configurations, typical system failures, etc.); *dynamic knowledge* about current system state, and *procedural knowledge* about how to carry out individual operator functions. Given the above model structure, all of this knowledge is interpreted by the model in either the network structure, i.e., its hierarchy, or by the state space, input and output alphabets associated with an individual network node. The model above provides a flexible framework for representing how an operator might organize system control functions. A limitation of the model is, however, the way in which the above knowledge is represented.

For the proposed system, the cognitive model of operator functions described above was established from gradually collected industrial experience (from discrete manufacturing industry) and, as practice indicated, built of high-level functions for planning/scheduling, monitoring, intervening and administering shop floor manufacturing activities during supervision. The model decomposes each high-level operator function into related subfunctions, tasks, and actions. Reactive scheduling is one of major scheduling functions included in the model.

However, the above model structure was not complete/suitable enough for a software design specification. To use it in the ISU as an effective assistant, it had to be provided with inferencing abilities. *Intent inferencing* provides the understanding properties of the model, as found in similar models indicated by, e.g., Geddes (1989). Ability to understand or infer the intentions of the human operator relies on a sound model of operator performance goals in the control of the system.

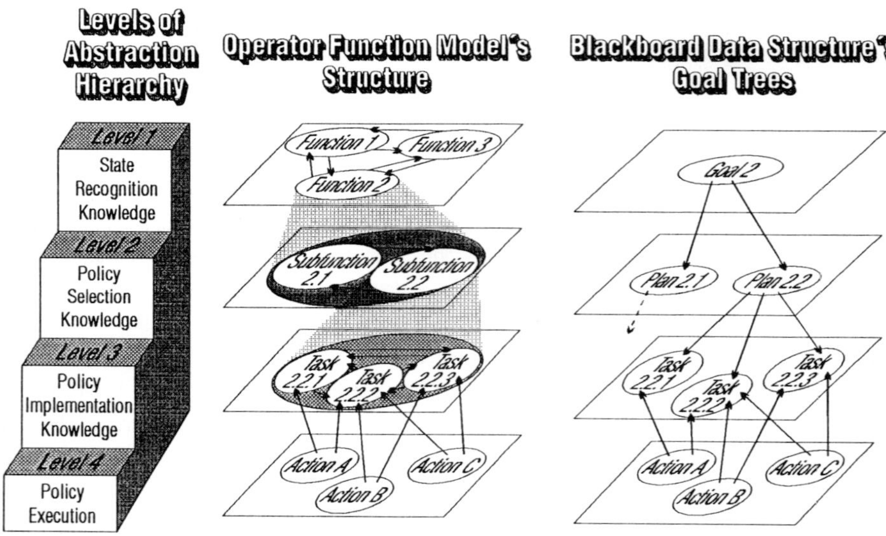

Figure 2. Abstraction hierarchy correspondence between Operator Function Model and Blackboard Data Structure

To provide the ISU of the system with intent inferencing ability, the cognitive model was completed with a goal-directed activity performance aspect. Thus, a currently instantiated operator function corresponds directly to a current operator goal. Currently active subfunctions associated with the active goals correspond directly to current operator plans, which in turn are carried out by different tasks and related actions (see Figure 2.). In complex dynamic systems, an operator often has multiple concurrent goals and multiple plans to achieve each goal. The dynamic nature of supervisory control in general and reactive scheduling problems in particular often necessitates goals which require prolonged and intermittent actions to maintain the controlled system in the desired state. All these considerations generate a need to provide the ISU with proper control capabilities, enabling it both to engage in system control and to interact with the operator in the mode of an intelligent assistant. To this end, the model is enhanced with control properties by an appropriate blackboard architecture, described in the next section.

4. BLACKBOARD MODEL FOR PROBLEM SOLVING IN THE PROPOSED SYSTEM

Management of the solution of shop floor control and implied reactive scheduling problems require the processing of large amounts of dynamically varying information/knowledge. From the design aspect of the intelligent supervisory control system, the *blackboard concept* application seems to be a viable choice. Among available software engineering techniques, blackboard systems possess the ability to deal with large amounts of diverse and incomplete information. From a knowledge engineering aspect, the blackboard framework of problem solving is a quite natural structure to represent the inferences and carry out the assessments of an intelligent supervisor. In the blackboard framework a hierarchically organized model of the problem is built (according to the knowledge abstraction hierarchy shown in Figure 2). It contains all that is currently known about the unfolding situation demanding a reactive scheduling decision and provides a context for on-going analysis on the nature of the disturbance (Nii 1986, Englemore *et al.* 1988). It is a proper medium for cooperation of the heterogeneous case and rule-based knowledge sources of problem solving in the system.

4.1 The blackboard architecture

The blackboard model of problem solving is generally composed of *three essentials* (Lalanda *et al.* 1992):
(1) The blackboard data structure, or the blackboard (BB),
(2) The independent knowledge-sources (KSs), and
(3) The control unit.

The blackboard can be characterized as a structured global database, a common problem solving area where the current best hypothesis is maintained and modified. The hypothesis is represented hierarchically, at various levels of abstraction, and evolves incrementally over time as new data become available or old data become obsolete. The BB is also the medium for all communications between the independent knowledge sources.

Solution elements are recorded on the BB by independent processes called *Knowledge Sources* (KSs), which respectively contain part of the domain specific knowledge required by the problem solving process and also knowledge for the control unit. Each change to the BB constitutes an event which can trigger one or more knowledge sources. KSs are responsible for posting, interpreting information on the blackboard, and contributing to problem solving.

A higher-level control structure called the *control unit*, or control KS is the kernel of the blackboard system. In the proposed system, it oversees the transfer of control from the user interface to the lower-level independent KSs (for case and rule-based reasoning in the system). Each KS has its own control structure for problem solving, and is designed to solve a specific problem in the system. Each KS stores and retrieves information from the BB. It provides means of cooperation between the knowledge sources which can be thought of as opportunistic since it discerns at any particular stage of the problem solving process which of the available knowledge sources is most suitable to contribute information. For this, three principal tasks are performed cyclically until a solution is successfully obtained or the process is halted by the user:

(i) the determination of which knowledge sources are most likely to provide useful information, i.e., the activation of the eligible knowledge sources;

(ii) the choice of one particular knowledge source susceptible to provide useful information;

(iii) the execution of the chosen knowledge source.

The ISU in the proposed system utilizes a blackboard data structure containing four levels of abstraction: goals, plans, tasks and actions (illustrated by Figure 2). The embedded model of operator functions defines what intentions (i.e., goal/plan/task structures) are associated with particular system triggering events and specifies how the current operator actions support such intentions (bottom up intent inferencing by the model). By this way, actions are immediately connected to whatever appropriate tasks exist on the blackboard at the time the actions are posted.

For example, if the operator, on request for decoupling a breakdown machine (for a given period of machine repair) from a scheduled manufacturing activity, could find an available identical substituting machine, he can simultaneously connect (free) substituting resources to all forthcoming operations queueing at the breakdown machine and scheduled predictively to the period of repair. By activating the preview simulation, he can reveal conflict sets of jobs (competing for the same substituting resource within the specific time frame) to be resolved by rule-based reasoning.

The operator function adequate to the dynamic decision situation and the problem encountered is activated by human. Then the model, given the system triggering event, defines subfunctions (plans), tasks and actions on which the operator should focus to solve the current problem.

The above hierarchical blackboard architecture is useful to make heterogeneous knowledge-sources (concerned with case-based reasoning/learning for storing user preferred repair plans for later use, or for retrieving old plans as solution alternatives from the case-base for a given problem instance, or with rule-based reasoning for constraints propagation/relaxation) to cooperate (see Figure 3).

The rule-based reasoning component in the system uses well-known techniques, such as forward and backward chaining for hard constraints propagation and soft constraints relaxation to resolve old and newly occurred conflict sets of job operations competing for the same finite resource within the specified time frame.

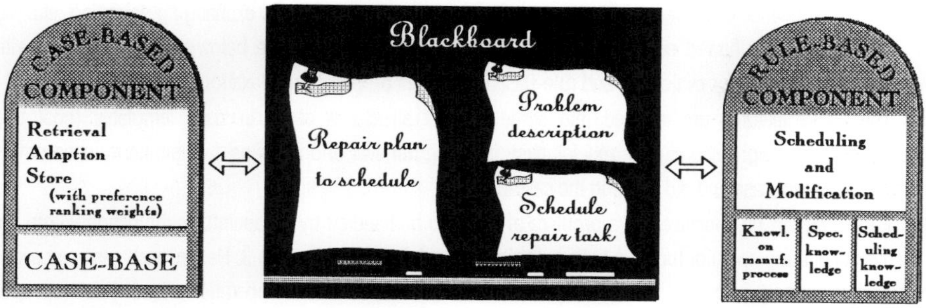

Figure 3. Interaction of case and rule-based components in problem solving

4.2 Case-based component (reasoning and learning)

Case-based learning/reasoning and rule-based components are to complement each other in the system. Due to the fact that knowledge is dynamic, it needs maintenance, i.e., the intelligent system should learn and adapt to changes in the environment and the manufacturing process. It has to discover and accumulate new ways of schedule repair, that is, learn from experience to reduce reactive scheduling effort by re-using learned skills and repair policies. The proposed system is designed to provide knowledge maintenance option for the system specialist via the learning component, as well as to enhance reactive scheduling capability of the shop supervisor.

The operation of CBR in the system essentially involves retrieving the most relevant past case(es) from the case-base, adapting them if necessary, and presenting the retrieved similar cases with successful repair policies, offering solution alternatives as well as implying preference ranking weights of human on effectiveness of alternative policies with respect to current performance criteria. Preference ranking weights are used to describe the desirability and acceptability of the alternatives being considered for selection.

The CBR component involves various concepts/techniques to organize information, and algorithms for indexing, searching for retrieval and assessing the similarity of cases. Generally, one way of representing cases is to view them as groups of features, with associated values. For case representation in the proposed system, we take into consideration a group of:

- global context features, describing averages and maximum deviations of the shop load, machines utilization and jobs lateness (where maximum deviations orient the desired direction of change in a repair situation), giving the number and relative location of bottlenecks in the process, etc.;
- local context features, such as current goal(s), current state of resources and that of jobs/orders (completion), type and effects of disturbance on goals/constraints/jobs; conflict sets; locally feasible repair

policies (sequence of some basic operations applicable locally for shifting, exchanging, inserting, etc. job operations in the flawed schedule, some of them are given in the Table below), complemented with preview simulation for evaluation and rule-based constraint propagation/relaxation if necessary.

The above features are mapped into some abstract structures of repair case templates (goal tree constructs with weight scores) utilized for case indexing/retrieval and allowing fast similarity assessment between a present case and old cases in the case-base.

Each case can be conceived as a problem of repair to be handled by an operator subfunction within the reactive scheduling operator function in the cognitive model detailed in Section 3. Hence, each case can evoke plans representing possible scenarios (given by sequences of basic operations and joint operator tasks/actions, e.g., to initiate a preview simulation run) that the repair process will encounter.

Change on the Fawled Schedule	**Change in the Process Causing Disturbances**
CANCEL:: job operation	(1) unavailability (lack, lateness) of needed resource ♣ (2) suspend a scheduled order to give advance to a new critical order ♥ (3) reroute a job by turning to a new technology alternative ♠
INSERT:: job operation	(1) expedit a critical new order ♥ (2) split a batch by involving another identical machine ♣ (3) involve substitute machine due to re-routing from breakdown machine ♣
EXTEND:: finish time of job operation	(1) add similar parts of rework (for scrapped parts) to a batch ♥ (2) adjust finish time due to technologically underestimated operation time (norm) ♠
REDUCE:: finish time of job operation	(1) reduce batch size with scrapped parts at previous operation ♥ (2) adjust finish time due to technologically overestimated operation time (norm) ♠
EXCHANGE:: two job operations	(1) lack of resources to the earlier and availability of resources to the later scheduled jobs ♣ (2) the later scheduled job has critical degree of lateness ♥
Legend: ♣ concerns RESOURCES ♠ concerns TECHNOLOGIES ♥ concerns ORDERS/JOBS	

Case matching is to be performed by the nearest neighbor algorithm (see, e.g., Jain and Dubes, 1988). This is achieved by attaching weights to each feature of the case-base and applying formulas to calculate the degree to which the stored case in the case-base matches the problem scenario of presented case. A case is similar to another based on the closeness of their score obtained through the weights.

The initial case-base is to incorporate a training set of stereotype disturbances (machine breakdown, expediting a critical new order, absence of operator, etc.), that cause delay in predicted finish times of scheduled job operations. The seed case-base is created to represent and organize information on disturbances and concerned repair policies by goal trees. Every case is indexed based on all its important features. The cases are then grouped (in a learning session by system specialist). Plans for repair are stored in a separate library to allow for re-use by different cases. In learning mode, new case structures are created by defining the features of a repair case template. With the case structures defined, new cases are entered through the above templates. A test function is provided to filter possible duplication in the case base. If a similar case exists in the case-base, the system specialist is allowed to adapt and generalize the case. Presented cases entered by human operator (user) with no match in the case-base are stored as unclassified new cases left for processing to the system specialist.

5. CONCLUSIONS—FURTHER RESEARCH TRENDS

Like majority of the systems in this domain, the proposed system has been in the model development and experimental phase, gradually proceeding toward building a prototype. This stage gives much more emphasis to the schedule quality and less to the resulting schedule efficiency. Efficiency issues are addressed in the system (and most of similar reactive scheduling developments) only to the extent that disruption to original schedule should be minimized by revising only those scheduling decisions that are found to be invalid, and this minimized disruption is correlated to reactive efficiency.

However, the more pragmatic problem of reacting to a change within a specified time frame has not been fully addressed. The pragmatics of timely reaction is a fundamental point in real-time interactive systems, such as the proposed supervisory control system, where the key words are: *sufficient and timely information.* In the proposed system, the human operator and the computerized system share supervisory responsibilities in a dynamic way, depending on the operator's current workload and the system state. Hence, the intelligent information display design (Szelke *et al.* 1993), enabling the operator to easily access the right information at the right time and in the right form, compatible with the human's current decision making needs, has been fully addressed in the system by the application of the cognitive model of operator functions within the ISU. The operator function model for the given supervisory control domain defines the knowledge needed not only for performing dynamic intent inferencing at control actions in general, but for designing the related dynamic information display pages in particular.

Further results in the context of reacting to a change within a specified time frame with provision of sufficient information are to be achieved in three ways in the proposed system. First, a gradual improvement of the system's knowledge-base/case-base by learning from experience can contribute to the provision of sufficient information to classified cases of reactive scheduling related decision situations. Hence, successful schedule repair policies can be readily re-used in similar cases, instead of rescheduling from scratch. Second, as timely information provision partly depends on the proper scheme of knowledge representation enabling fast information retrieval from the case-base, we store indexed cases by goal trees using frame/object representation, to ease storage/retrieval. Third, real-time capability of problem solving is to be built into the system's blackboard control architecture. The proposed system is being developed with a sophisticated control unit with a structure giving the control to the most important procedure (knowledge source) execution within specified time windows, while allowing the interruption of the other's execution. The ability to handle interrupts is to be accomplished by storing the state of the reasoning system (current evaluation of inferences, goals, and other related facts attending and serving the time critical real-time interrupt) and re-entering the state taking into account changes during the interrupt, which could have modified the original state.

Through the designed architecture the system is capable of providing the human with solution alternatives to problem instances and learning preferred choices of the human (operator) from feasible alternatives, within a more limited amount of time than manually possible. To operate under severe real-time constraints, the inherent opportunism of blackboard systems must be balanced by the above special control mechanism in the control unit of the blackboard architecture to satisfy the need for predictability in response time.

The problem of this real-time regime and the case and rule-based KSs interaction in the problem solving process are subjects of further research. The above problems, together with the utilization of the Object Oriented Programming (OOP) paradigm at the system implementation (since blackboard models lend themselves easily to be implemented by using the OOP paradigm), represents a significant amount of forthcoming R&D work.

Further longer term research is concerned with building a multi-agent perspective into the system, i.e., to involve all the operators in the control and decision making processes (thus into the reactive scheduling decisions), including also the operators concerned with the auxiliary processes of manufacturing (operators for preparatory engineering, quality assurance, preventive maintenance, etc. sections), that is, to provide the system with a cooperative intelligent agent architecture. These systems demand a framework (Smith 1993, Hayes-Roth 1990) for prioritizing and managing pending conflicts among the contributing agents by, e.g., negotiations.

All the above outlined development perspectives are related to ongoing research in the domain supported by more and more advanced AI tools.

Acknowledgement

Research has been supported by a grant (No. T 4119) of the Hungarian Scientific Research Founds in the period of 1991–1994.

REFERENCES

ATABAKHSH, H., 1991, A survey of constraint based scheduling systems using an artificial intelligence approach. *Artificial Intelligence in Engineering*, 6 (2), 58–73.

BENSANA, E., BEL, G., AND DUBOIS, D., 1988, OPAL : A multi-knowledge-based system for industrial job-shop scheduling. *International Journal of Production Research*, 26 (5), 795–819.

BERRY, P. M., 1990, Resolving conflicting objectives in factory scheduling. *Proceedings International Conference on Expert Planning Systems*, (Institution of Electrical Engineers, Brighton, UK), 16–21.

BIENNER, F., BELSON, G., AND FAVREL, J., 1992, Utilisation de reseaux neuronaux pour le pilotage d'ateliers flexibles. *Proceedings of 5th International Conference on Neural Networks and their Applications*, (Nimes, France).

BRUNO, G., ELIA, A., AND LAFACE, P., 1986, A rule-based system to schedule production. *IEEE Computer*, (6) 32–41.

ELLEBY, P., FARGHER, H. E., AND ADDIS, T. R., 1988, Reactive constraint-based job-shop scheduling. In: Oliff, M. D. (Ed.), *Expert Systems and Intelligent Manufacturing*, (North-Holland, New York), 1–10.

ENGELMORE, R. S., MORGAN, A., AND NII, H. P., 1988, Introduction. In: *Blackboard Systems*, Eds.: Engelmore, R. and Morgan, A., (Reading, MA: Addison-Wesley), 1–24.

FOX M. S., 1983, Constraint Directed Search: a case study in job shop scheduling. *Ph.D. Thesis*, (Carnegie Mellon University, Computer Science Department, Pittsburg, USA).

GEDDES, N. D., 1989, *Understanding intentions of human operators in complex systems.* Ph.D. dissertation, (School Ind. Syst. Eng., Georgia Inst. Technol., Atalanta).

HOONG, L. W., YEO, K. T., AND SIM, S. K., 1991, Application of neural networks in a job-shop environment. *Proceedings International Conference on Computer Integrated Manufacturing ICCIM'91*, "Manufacturing Enterprises of the 21st Century", (World Scientific, Singapore), 559–562.

JAIN, A. K. AND DUBES, R. C., 1988, *Algorithms for Clustering Data.* In: *Advanced Reference Series*, (Prentice Hall, Englewood Cliffs, New Jersey, 1988).

KEMPF, K., RUSSEL, B., SIDHU, S., AND BARETT, S., 1991, AI-based schedulers in manufacturing practice. *AI Magazine*, 11 (5), 46–56.

KERR, R. M., AND KIBIRA, D., 1993, *Simulation and learning in knowledge based scheduling.* (School of Mechanical and Manufacturing Engineering, University of New South Wales).

KERR, R. M., AND WALKER, R. N., 1989, A job shop scheduling system based on fuzzy arithmetic. *Proceedings 3rd Int. Conf. on Expert Systems and the Leading Edge in Production and Operations Management,* (Hilton Head Island) 433–450.

KERR, R. M., 1991, *Knowledge Based Manufacturing Management,* (Addison-Wesley, Sydney).

KHAW, J., SIONG, L. B., LIM, L., YONG, D., JUI, S. K., AND FANG, L. C., 1991, Shop floor scheduling using a three-dimensional neural network model. *Proceedings International Conference Computer Integrated Manufacturing, ICCIM'91,* "Manufacturing Enterprises of the 21st Century", (World Scientific, Singapore), 563–566.

KOLONDER, J., SIMPSON, R., AND SYCARA, K. P., 1985, A process of case-based reasoning in problem solving. *Proceedings the Ninth International Joint Conference on Artificial Intelligence,* (Los Angeles, USA), 284–290.

LAFFEY, T. J., COX, P. A., SCHMIDT, J. L., KAO, S. M., AND READ, J. Y., 1988, Real-time knowledge-based systems. *AI Magazine,* (9), 27–45.

LALANDA, PH., CHARPILLET F., AND HATON, J. P., 1992, A real-time blackboard-based architecture, *Proceedings of 10th European Conference on Artificial Intelligence—ECAI' 92,* (ed.) Neumann, B., (John Wiley & Sons Ltd., 1992), 262–266.

MITCHELL, C. M., 1988, Supervisory control: Human information processing in manufacturing systems. In: *Concise Encyclopedia of Information Processing in Systems and Organizations,* (Pergamon Press, New York).

MITCHELL, C. M., AND MILLER, R. A., 1986, A discrete control model of operator function: A methodology for Information Display Design. *IEEE Transactions on Systems, Man and Cybernetics,* Vol. SMC-16, 3 May/June 1986, 353–369.

NII, H., 1986, Blackboard systems, *AI Magazine,* 7 (2), June 1986.

NORONHA, S. J., AND SARMA, V. V. S., 1991, Knowledge-based approaches for scheduling problems: A survey. *IEEE Transactions on Knowledge and Data Engineering,* 3 (2), 160–171.

OW, P. S., SMITH, S. F., TIRIEZ, A., 1988, Reactive Plan Revision, *Proceedings Seventh National Conference on Artificial Intelligence,* (American Association of Artificial Intelligence, St. Paul, USA), 77–82.

RASMUSSEN, J., 1990, A model for the design of computer integrated manufacturing systems: identification of information requirements of decision makers. *Int. Journ. of Industrial Ergonomics,* Elsevier, 5, 5–16.

ROUSE, W. B., GEDDES, N. D., AND CURRY, R. E., 1987, An architecture for intelligent interfaces: Outline of an approach to supporting operators of complex systems. *Human-Computer Interaction,* 3 (2).

SADEH, N., AND FOX, M. S., 1990, Variable and value ordering heuristics for activity-based job-shop. *Proceedings of the Fourth International Conference on Expert Systems in Production and Operations Management,* (Hilton Head Island, USA), 134–144.

SARIN, S. C., AND SALGAME, R., 1989, A knowledge-based system approach to dynamic scheduling, In: Kusiak, A. (Ed.), *Knowledge-based Systems in Manufacturing,* (Taylor & Francis, London), 173–203.

SHERIDAN, T. B., Supervisory Control, in: Salvendy, G., (Ed), *Handbook of Human Factors*, [Chapter 9.6], New York: John Wiley and Sons, 1987.

SMITH, S. F., 1987, A constraint-based framework for reactive management of factory schedules, In: Oliff, M. (Ed.), *Proceedings First International Conference on Expert Systems and the Leading Edge in Production Management*, (Charleston, USA), 349–366.

SMITH, S. F., OW, P. S., POTVIN, J. Y., MUSCETTOLA, N., AND MATTHYS, D., 1990, An integrated framework for generating and revising factory schedules. *Journal of the Operational Research Society*, 41 (6), 539–552.

SMITH, S. F., 1992, Knowledge-based production management: approaches, results and prospects. *International Journal of Production Planning and Control*, 3 (4), 350–380.

SMITH, S. F., 1993, OPIS: A methodology and architecture for reactive scheduling. In: Fox, M., and Zweben, M. (Eds.), *Intelligent Scheduling*, (Morgan Kaufmann Publishers).

SZELKE, E., AND KERR, R. M., 1993, Knowledge-based reactive scheduling, *International Journal of Production Planning and Control*, 5 (1), (In print, Taylor & Francis, London).

SZELKE, E., AND MARKUS, G., 1993, Intelligent interface design to supervisory control of dynamic discrete-part manufacturing processes. *Proceedings of the IFIP Working Conference on Interfaces in Industrial Systems for Production and Engineering*, (IFIP Transactions B-10, Elsevier Science Publishers, Amsterdam), 159–173.

SZELKE, E., MESZAROS, I., Intelligent and adaptive Control of FMS contributing to CIM trend, In: *Proceedings 11th IFAC World Congress on Automatic Control in the Service of the Mankind*, Tallin, 1990, Eds.: Utkin, V., Jaksoo, U., (Pergamon Press, IFAC, 1991) Vol. V., 31–36.

SYCARA, K. P., AND MIYASHITA, K., 1993, Adaptive schedule repair. *Preprints of the IFIP Workshop on Knowledge-Based Reactive Scheduling*, (IFIP/Hellenic Management Association, Athens, 1993), 105–127.

WOODS D. D., 1986, Cognitive Technologies: The design of joint human-machine cognitive systems. *The AI Magazine*, 86–92.

WOODS, D. D., ROTH E. M., AND BENNETT, K., 1990, Explorations in joint human-machine cognitive systems. In: *Cognition, Computing, and Cooperation*,(Eds.) S. Robertson, A. Zachary and J. B. Black, Norwood, N. J. Ablex, 1990.

YUAN, Y., AND WU, Z., 1991, Algorithm of fuzzy dynamic programming in AGV scheduling. *Proceedings International Conference on Computer Integrated Manufacturing*, ICCIM'91, "Manufacturing Enterprises of the 21st Century", (World Scientific, Singapore), 405–408.

Knowledge-Based Reactive Scheduling (B-15)
E. Szelke and R.M. Kerr (Editors)
Elsevier Science B.V. (North-Holland)

A Predictive and Reactive Scheduling Tool Kit for Repetitive Manufacturing

M.M.C. Tam[a], *D.H.L. Choi*[b], *W.W.C. Chung*[c], *T.C.E. Cheng*[d], *and P.P.K. Chiu*[e]

[a] *Department of Computing, Hong Kong Polytechnic, Hong Kong*
[b] *Wing Sang Bakelite Electrical Manufacturing Company Limited, Hong Kong*
[c] *Department of Manufacturing Engineering, Hong Kong Polytechnic, Hong Kong*
[d] *Department of Management, Hong Kong Polytechnic, Hong Kong*
[e] *Department of Electronic Engineering, Hong Kong Polytechnic, Hong Kong*

ABSTRACT

This paper describes a Predictive and Reactive Scheduling Tool Kit. The concept of a tool kit approach is to integrate some basic and useful tools in production planning and control operations. The proposed tool kit, at its initial stage of development, is composed of three basic tools. The way that these basic tools are integrated is demonstrated by a case study in a local repetitive manufacturer. In the case organization, the resulted tool kit is used to support a software process which is intended to provide predictive and reactive scheduling support to assist material planners in production planning and control. The advantages and shortcomings of the tool kit approach in providing decision support is empirically examined. Based on the empirical study, further enhancement in predictive and reactive scheduling is suggested.

Keyword codes: I.2.4; K.6.1; H.4.3
Keywords: Knowledge-Based, User-oriented DSS, Common shared window

1. INTRODUCTION

1.1 Background and Motive

In the past few years, Hong Kong Polytechnic was approached by local manufacturers who were looking for academic support to exploit effective use of information technology for production planning and control, typical examples were inventory and order management, production scheduling, etc. The primary aims of these projects were to demonstrate the need for coherency in research and practice; and to promote creative use of information technology to assist production management. Typically, these projects involved the design and development of Decision Support Systems (DSS) to aid production executives in operation management. They were encouraged to develop their user-oriented DSS themselves. Much empirical evidence showed that this kind of user-oriented development was beneficial in terms of mastery, data integrity and disciplines. In order to carry the idea of user-oriented development a step further, the concept of a tool kit approach was introduced. The idea of the tool kit approach was that a set of tools could be derived and used in a wide range of applications. In order to test this idea, joint research projects with local manufacturers were organized and system developments were carried out in host companies.

The design and development of DSS is naturally oriented towards the context of repetitive Manufacturing, as a number of manufacturers in Hong Kong are already running their business in this mode of operation. In fact, there are many unique operational characteristics in such a context. It is also a result of them that we begin to realize that application of predictive and reactive scheduling can be an effective production planning and control strategy to assist production executives.

The remainder of the paper is organized as follows: Section one is an introduction which covers some terms and definitions that are used in this paper. Section two conceptualizes the production planning and control operations in the repetitive manufacturing. Section three illustrates the design of the tool kit for this context. Section four is a case study which demonstrates the implementation of the tool kit in a real life setting, its shortcomings and advantages are identified as well. Finally, section five concludes this paper.

1.2 Terms and Definition

Scheduling is a knowledge intensive activity in terms of both domain and general scheduling knowledge. The domain knowledge that will be covered is related to repetitive manufacturing. In addition, some general concepts in relation to the tools will be introduced later are also defined in order to avoid confusion that may arise.

Repetitive Manufacturing: The term is defined as the fabrication, machining, assembly and testing of standardized units produced in volume, or as "anonymous" sub-assemblies assembled-to-order in volume from standard options [1,2]. Typical products include electrical appliances, civil electronics, clocks and watches. On account of the need for high volume production, production lines tend to be dedicated, at least as far as assembly operations are concerned. In most cases, material is assumed to be in continuous consumption (i.e., all upstream production units are required to produce enough to sustain the operations of down stream production units at any time). In other words, timely planning, monitoring and control of material flow is an important element for effective production management.

Assemble-to-Order (ATO) Policy: Anonymous products (mainly sub-assemblies, WIP) is a key feature of assemble-to-order manufacturing policy. Anonymous products are produced in advance and made available for allocation when a new order arrives. They are designed with an implicit goal that they can be used in many different product lines, and therefore they are less prone to obsolescence. There are several practical rules which could be used together with these design tactics in order to enhance the logistic performance of repetitive manufacturing. As far as material management is concerned, these rules are embedded in the following practices:

- keeping critical anonymous products as strategic buffers against uncertainty
- using these strategic inventories as stored capacity
- planning anonymous products in aggregate amounts which justify volumetric production

- using a hybrid procurement policy: purchase-to-stock and then assemble-to-order to shorten manufacturing lead time

"Anonymous product" manufacturing allows a manufacturer to delay early commitment to a particular customer, and hence, retain a higher level of flexibility to change. Furthermore, from a statistical point of view, aggregate planning of anonymous products helps to reduce prediction errors in material requirement.

Predictive and reactive scheduling: Predictive and reactive scheduling has been demonstrated to be an useful concept for modelling production management information systems [5,6]. The basic principles involved are (1) planning is done in advance using most of the available information, and (2) reacting to the earlier actions, as a consequence of earlier prediction, and considering new information as it becomes available.

In that regard, predictive and reactive scheduling takes a hybrid approach towards scheduling. The planning, similar to traditional philosophies of scheduling, is done in advance and is prepared to be modified (i.e. perform schedule repair) to cope with emerging situations. Reactive schedule repair is not feasible unless there is certain degree of freedom in making use of the opportunity (i.e. it must not be constrained). In fact, there are several structural design features typically embedded in ATO repetitive manufacturing which open up opportunities for schedule repair. This is also a reason why schedule repair is a distinct feature in repetitive manufacturing. On account of the higher level of standardization as a result of anonymous production, there exists a higher level of interchangeability between product lines and hence a larger degree of freedom in swapping the product or capacity. Also, swapping may take place at several levels in the product structure (i.e. BOM). For example, interchangeability exists at any level of the BOM (e.g. the raw materials, the final assembly, the order). The general principle here is that the less of a commitment in terms of processing towards a particular customer, the more flexibility is retained.

2. PPC IN REPETITIVE MANUFACTURING

Due to the uncertainty in forecasting the mix and volume of customer orders, production planning and control for it is difficult to produce the right materials and to install the right capacities beforehand. It is then even more difficult in situations where the material supply is not perfect, when customers change their requirements and engineers change the specifications of the product all at the same time. It is noted that assemble-to-order [3] manufacturing policies could help facilitate production control to cope with these challenges. However, effective execution of assemble-to-order in practice usually necessitates more coordination, and as a result, requires more information and decision support.

At an analytical level of examination, the context diagram of a production system of repetitive manufacturing can be conceptualized as shown in Figure 1. In this diagram, a production system is represented as a network of production units separated by inventory points. A production unit is regarded as being self-contained, having a number of capacity resources at its disposal. Material flow control would be required for the coordination of the production unit. The assumption concerning about the relationship between production unit and material flow control implies that apart from the operational relationship between input and output, the internal structure of the production unit is not relevant as far as it meet some aggregate agreements (e.g. lead-time, output, yield rate, etc). In other words, material flow control does not need to know the details of operations for manufacturing specific items, the actual availability of capacity type, the sequencing rule that are used, or the exact location of the work orders on the shop floor. Only for those variables which set the operational limits, material control must know the internal state of the production unit.

Given this simplification, the scheduling tasks could be reasonably reduced to one of coordinating material flow, wherein the operational characteristics of the production unit are regarded as parameters to be considered in scheduling. An activity based model of such context is summarized in Table 1.

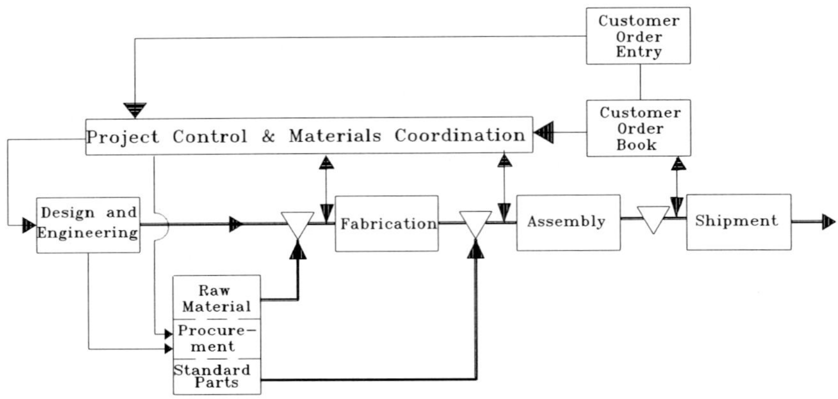

Figure 1. Context Diagram of Repetitive Manufacturing

3. DECISION SUPPORT WORKBENCH PROTOTYPE

The decision support workbench is tested by synthesizing a set of tools to provide the required predictive and reactive decision support and coordinate decision making.

3.1 Theoretical Foundation

At the time of our project work, three basic tools, namely, cumulative in / out modelling, recursive BOM querying techniques and Petri-Net modelling were experimentally developed and then field tested in the case organization.

An Activity Base Model of a Predictive & Reactive Scheduling Framework	
PREDICTIVE ACTIVITIES	Given that there are aggregate agreements between sales, purchasing and production in respect with service performance, stability and reliability of required delivery patterns, total accepted demand, etc., it is possible and desirable for sales to negotiate with the customers and formulate the sales plan autonomously. Furthermore, as a consequence of the dedicated production lines which implicitly mean that the workload control is largely confined, the average lead-time preset and the delivery performance fixed, it is also possible for sales to consider these logistics performance factors together in the estimation of delivery plan meanwhile formulating the sales plan. Similarly, production and purchasing could also formulate their production and procurement plan autonomously.
REACTIVE ACTIVITIES	However, more and larger safety factors (e.g. over capacity, larger safety stock, later delivery date) are often required which suggest that any decision of this kind would hardly be an optimal one. It is reasonable to expect that further improvements (may be the next level of sub-optimization) could be made if the actual manufacturing status is used whenever it becomes available. There are strong arguments as well as evidences that operational coordination (e.g. work order release control, shop floor monitoring) could be of vital importance in that regard [4]. For example, proper work order release decisions could contribute to improvements in logistics performance [5]. Higher level of operational coordination naturally implies that production planner and controller need to cooperate with each others. In most circumstances, they have to negotiate with outside agents such as customers and vendors. Their genuine practices, in most circumstances, are attributed to be tactful to excavate the highest degree of flexibility which would otherwise be impossible.

Table 1. An Activity Based Model of Predictive and Reactive Scheduling

Petri-Net modelling has been used extensively in analysing dynamic systems (e.g. manufacturing). Some of the advantages of using Petri-Net are formality, structure, and also one of its advantage lies in its simplicity and visual modelling of manufacturing systems. In this project, we are going to enhance these two attributes by extending a visual echelon modelling and integrating it with the traditional model of BOM processing. The advantage of the kind of integration in a repetitive manufacturing context can be understood as follows:

In the simplest case where there is little interchangeability, single machine, the situations of discrete product repetitive manufacturing is similar to that of a job shop context. In such a job-shop production environment, effective control of an order queue length could make significant improvements. Petri-Net can be an useful tool in that regard. However, since the buffer can be reasonably tolerated and expediting can be reasonably loosened, the relative importance of queue control can be less significant in repetitive manufacturing. The centre of interest is on a collective measure of throughput as long as there is a sufficient buffer. Petri-Net is less effective, in its conventional sense of token passing, in modelling throughput dynamics. In the proposed design, the concept of token passing is extended as a time series variable and it is shown as an echelon to enhance its visual impacts.

It is not unusual to have a high level of substitution in terms of product interchange and order interchange. The implication on the system to be modelled by a Petri-Net is that there are many possibilities. In addition, change in the dispatching routine could also have an impact on the structure of the net. This kind of flexibility in demand could be difficult to manage. In the proposed design, the knowledge of the net is encapsulated by the techniques similar to those used in the structure of a BOM. An auxiliary tool is provided in maintaining the integrity as well as the flexibility.

The software process to be supported is shown in Figure 2. The functionality of the first and the third block together is in several ways similar to that of traditional MRP. The major functionality that intended to be covered is time-phased requirement calculations: requirements at every level of the BOM are consolidated, offset by lead time, and time phased which are lastly translated into firm orders and schedules. The additional features complemented the shortcomings of MRP in a repetitive manufacturing context.

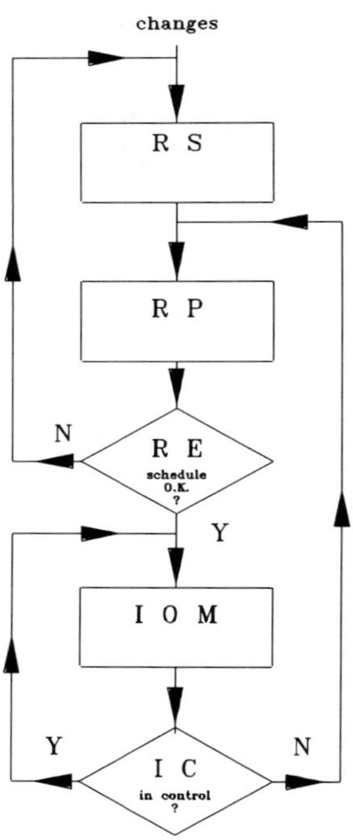

Requirements Scheduler (RS) Actual time phased material requirements need to be updated promptly within each operational cycle. In each cycle, the latest information is solicited which is then used to adjust the requirements for the nearest periods and to define a trend for the more remote ones. Suitable software interface is required for entry of order-booking.

Requirements Profiler (RP) The resource profiler is used to highlight the shortcomings of a schedule. It takes into account of orders already booked, the work order backlog and the replenishment schedule. The scheduler could use the RP to project all these order bookings into the throughput diagram to foresee the consequences of these anticipated transactions.

Requirements Evaluator (RE) The core of RE is a constraint directed search algorithm. A two step objective function is defined for the search, where the first step is to identify potential material shortage and then the second step to try to eliminate it. The search tree is produced by exploding the schedule of a customer order and work order to its lowest level of resource commitment. The explosion is defined by the product structure, i.e. BOM. The already commitment of resources is modelled as the constrain of the search algorithm. Entering new material replenishment orders or retracting existing orders modifies the commitment variables and therefore represents possible relaxation in the constraint and hence extended the search domain. Dispatching routines are also considered which provide the necessary reactive schedule repair.

Inventory and Order Manager (IOM) The basic transactions appear in the flow balance model form the core of the inventory order management model. It provides an overall picture of the actual situations on the basis of inventory levels, and together with other order bookings to project future trends.

Figure 2. A Flow Chart of the Predictive and Reactive Scheduling Software Process

3.2 Application Level of Development

An alternative Predictive and Reactive Scheduling framework [6] that has been demonstrated to be an useful concept for modelling production management information systems is shown in Figure 3.

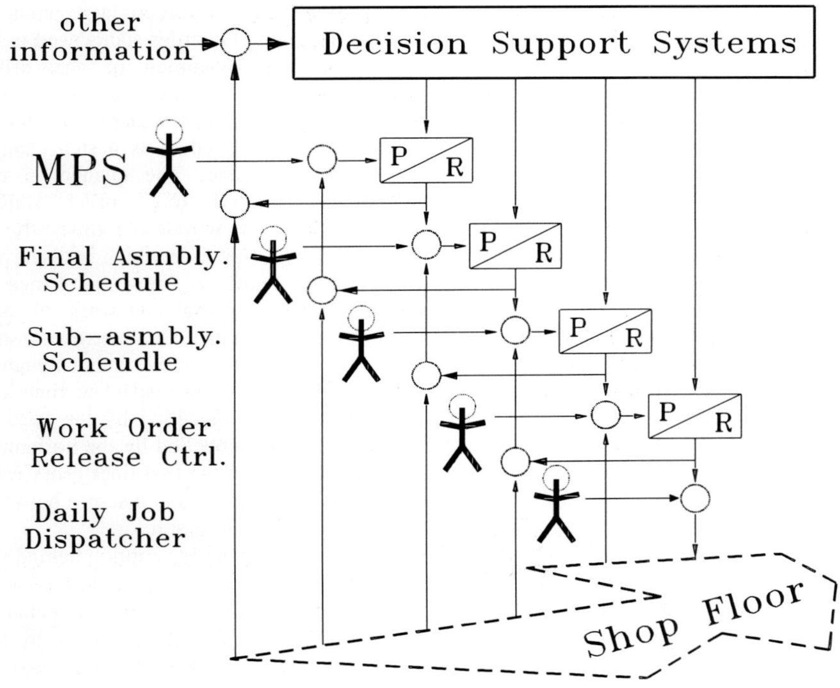

Figure 3. A Predictive and Reactive Scheduling Hierarchical Architecture

In theoretical terms, any agent in the hierarchy is assumed to be composed of two parts: a Predictive element and a Reactive element, which together is a P/R agent. In principle, all agents perform as an autonomous unit. Upper level agents give predictive guidance which provides an additional bias in making control decisions for lower level agents. The elements behave as R-agents and coordinate themselves in a hierarchical way when they are coordinated for control, and yet each element can still function as an independent P/R agent when necessary. It is not strictly hierarchical because information feeds back to different levels of the hierarchy directly without traversing the intermediate nodes. The lowest decision making level reacts to events first, and remedies the situations as far as possible before necessitating supports from immediate upper level agents.

3.2.1 Cumulative In / Out Curve

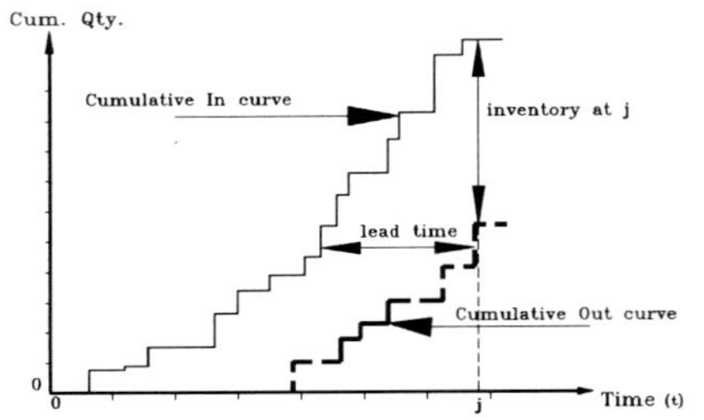

Figure 4. Cumulative In / Out curve in modelling material movements

As every order cycle involves transactions to both open and close an order, large amounts of data concerning the order status need to be monitored. The dynamics of the order transactions could be modelled by a linear step function where each variable of interest in the transaction is taken as a step change with respect to time. The change in the variable of interest (e.g. order status) therefore is a function of time. It is represented by a time line which is displayed graphically in Figure 4. Here, the change of incoming and outgoing quantities of any particular item are of interest for order status monitoring and tracking. The interrelationship between the two lines represents the dynamics of interdependency. The two major dynamic patterns of interest are: change in material shortage and surplus with respect to time and change of item lead time with respect to time. The former is modelled by the vertical separation between lines (the stock on hand), while the latter is modelled by the horizontal separation between lines (the lead time relationship). Whenever these two lines cross each other, it indicates material shortage. This concept will be clearer if the gap distance between the two lines is represented as the balance and shown on a separate graph.

However, returning to the hierarchical framework in Figure 3, the assumption behind the partitioning framework is that a high level of local autonomy can be maintained to their affairs in balance. This assumption is not far from reality given that suitable tools to enhance production visibility are available. Representation techniques, in this case, (shown by the graphical representation in Figure 4) can be an effective visual aid. The resulting diagram is regarded as the Cumulative In / Out curve. These curves can be used to represent the logistical goal for manufacturing control.

3.2.2 BOM processing

In the above modelling process, PPC in a repetitive manufacturing is considered to be a series of interlinked planning and control loops. The specific visual modelling technique offers capabilities for monitoring and control of individual units. Their integration is through establishment of inter-loop linkages at various levels (e.g., the item, BOM, order, fund flow, project, and department). This integration is assisted by a pre-planed decision hierarchy in the BOM processing. The generic use of BOM processing can be categorized into (1) methods for planning future operations and (2) communicating scenarios. With respect to this categorization, each PPC loop could be considered to be similar to the extent that it includes the following common activities:

Planning Future Operations: Production requirements and schedules are prepared for each level, based on committed or forecasted requirements. For finished products, the higher level requirement will be the customer order; however, for purchased parts, the higher level requirement will be the assembly schedule. Here, the BOM is used to explode the assembly into its subassemblies or components. An important technique for constructing the BOM for decision support is to embed preplanned decision points in the BOM processing. Some typical examples include: inclusion of lead time offset for time-phasing material planning, rolling up cost figures, and scheduling engineering changes, etc. [9].

Communicating Scenarios: Next, the part requirements are processed against the stock status and each part analyzed for availability. Another kind of preplanned decision procedure can be effective at this point. One kind of preplanned decision here is to allow for order transaction and analysis (e.g., actual or logical allocation of materials or queuing of order backlog). The work or purchase request is then prepared for those parts need to be replenished. PPC prepares the work orders, collects the relevant work documents and then releases them according to the schedule.

Monitoring Operations: Data collected about the stock before and after each production unit is available to keep track of the progress of an order. Once the activity cycles for every PPC loop and their inter-linkages have been identified, it is possible to design the necessary data models, system outputs, control parameters, and system inputs. Such design considerations are covered in terms of coordinated decision making and management of inventory and information.

Transitional BOM processing represented by a Transitive Closure Logic is applied in the formulation of recursive property of BOM. The transitive closure rule for recursive BOM processing (level by level explosion) is defined as follows:

R(X,Y):-A(X,Z),R(Z,Y),

where X is a Job Number which maps to an assembly or a sub-part, and Y is the set of all children of all levels of that assembly or sub-part; and the exit rule is defined as:

R(X,Y):-A(Y,Y),

which means the search of children terminates when X=Y.

3.2.3 Petri-Net Procedural Knowledge Encapsulation

Understanding the procedure would be an important requisite towards knowledge-based development. In our study, it is assisted by a variant of the Petri-Net model which is adapted from Sartori, L.G.[7], an example is shown in Figure 5.

In this model, logical variables are represented by circles and numerical variables are represented by capsules. These techniques could be useful in identifying the control point in a manufacturing system. Also, procedural knowledge can be derived from the model and translated in a form of production rule, for example:

IF boolean ORDER is TRUE (open), THEN proceed ISSUE.
IF numeric Qty_on_hand < Qty_Safety_stock, THEN trigger on REPLENISHMENT

This type of knowledge can be processed by a knowledge-based system with little difficulty, so a knowledge-based system prototype used Object-Oriented techniques is developed to assist this knowledge acquisition work.

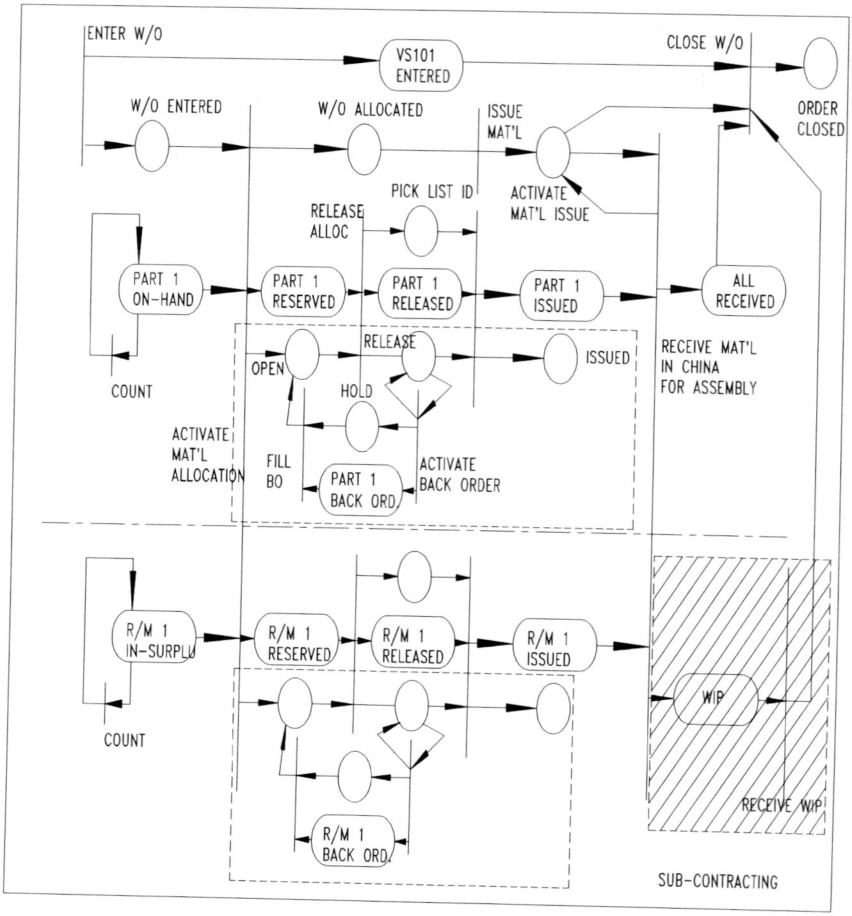

Figure 5. An Example of use of Petri-Net Model Procedure

4. Case Study

Wing Sang Electrical Manufacturing Ltd. (hereafter referred as WSM) was established as a local original equipment manufacturer (OEM) in 1960. In this company, a "Predictive and Reactive Scheduling Tool Kit" had been recently developed by the line staff under a software development environment. It simply consisted of a relational database, a graphic generator, and a report writer working under a network environment. Predictive and reactive knowledge, and use of the DSS tools were transferred directly from the staff (developers) to the staff (target users) in the production office. Thus, a new information system was developed. It was used to provide a common shared window in the production office. An example of which is elaborated in Figure 6. All the up-to-date transaction data could be visualized, and earlier responsive action could be taken to deal with forthcoming schedule discrepancies. For instance, if the purchaser views that the Cumulated In Curve overlapped with the Cumulated Out Curve, he/she may check up the outstanding purchase orders from his/her order book against the material requirement shown in the schedule and takes the responsive actions (e.g., call up vendor for urgent delivery, salvage over-designed resources, inform staff in production office to use a substitution, etc.) autonomously in order to prevent a temporary material shortage. The knowledge-based tool kit was first implemented on July 1 1991 and the effort was focused on ATO assemblies.

Figure 6.a shows the "Stock Balance graph for '900-100-10-XX10' from the period of 91.07.01 to 92.03.31". This is a commonly used sub-assembly at WSM. It is continuously assembled in the factory, stored in the warehouse, and then released as a lot for final assembly when customer orders are received.

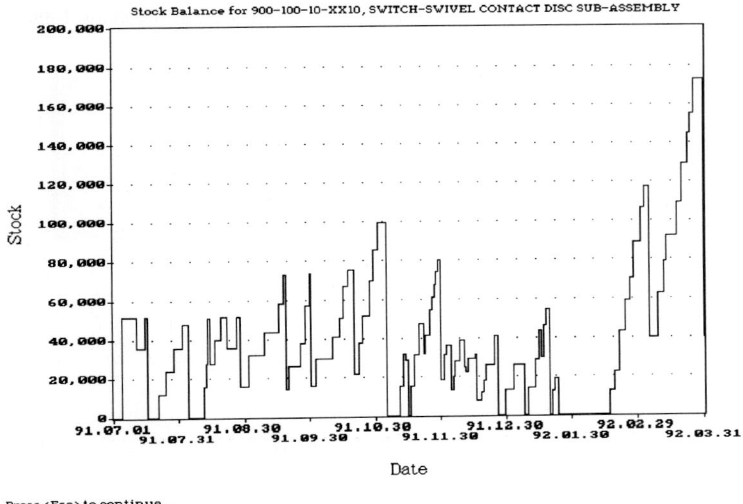

Figure 6a.

Figure 6.b shows the "Stock Balance for '900-100-10-XX10' from 92.01.01 to 92.03.31". There is no inventory and no production record during the period of 92.01.21 to 92.02.30 as this period is the Chinese New Year holiday.

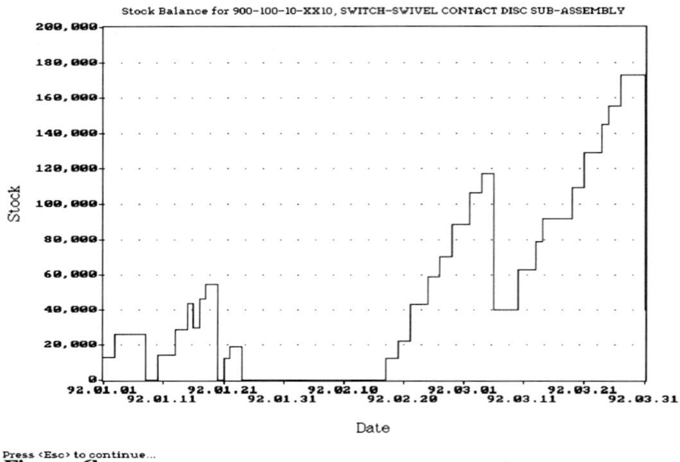

Figure 6b.

Figure 6.c is the "Stock In / Out for '900-100-10-XX10' from 92.01.01 to 92.03.31 (include data)". The balance graph in Figure 6.b is derived from the subtraction of the In / Out graph in this figure. Both the graph and transaction data can be viewed at the same time.

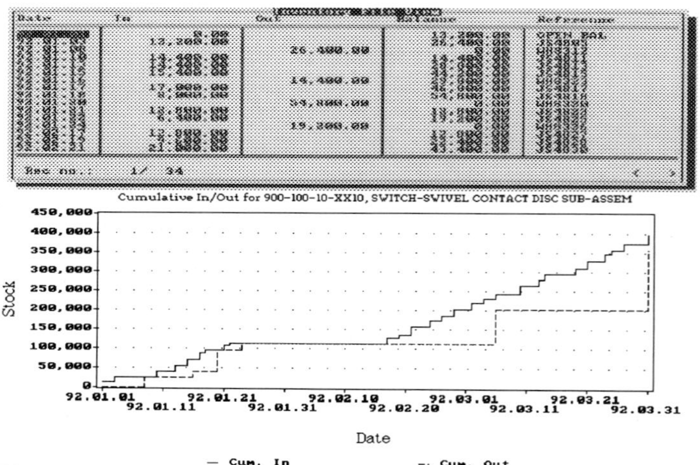

Figure 6c.

Figure 6.d show the "Stock In/Out for '900-100-10-XX10' from 92.01.01 to 92.03.31 (exclude data)". Production volume grows after Chinese New Year because more products have to be shipped to the customers before Easter and Mothers' Day.

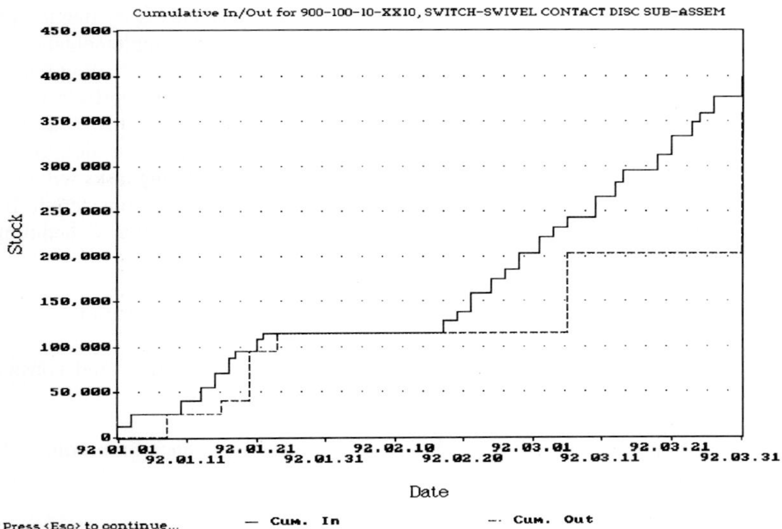

Figure 6d.

The implementation results demonstrate that the tool kit is successful as appraised by the production staff and management team in the company. The material in or out status is kept active, which provides timely information for decision making. Furthermore, implementation of the tool kit improves on inter- and intra-departmental coordination, the detection of bottlenecks for taking earlier reactive action. It serves as a set of predictive and reactive scheduling tools that is applicable to various planning levels, and also as a distributed resources planning for logistics support of the operations in remote production plant in China. Data which is previously not easily accessible is now available to the production staff. In other words, they are able to coordinate their re-scheduling efforts particularly in accommodating for short term variations (e.g.: customer order change, work order priority change or temporary material shortage, etc.). However, the methods to deal with the schedule discrepancies are different from case to case and the manufacturing environment always changes. The line staff now focus their re-scheduling efforts in the re-design of operation procedures and establishment of process parameters in the production office using Petri-Net Modelling [8] to improve their response to discrepancies. Provided that these procedures and parameters are rationalized, knowledge-based techniques could be used more effectively.

5. Conclusion

Based on the projection of future utilizations of manufacturing resources, scheduling decisions are made which involve early resources commitments in advance. These resource allocation decisions are intricate with each other and together with a lot of stochastic uncertainties. One effective way to overcome scheduling discrepancies is by making proper use of local autonomy in which all the tasks needed to coordinate the implementation of the schedule (including scheduling sub-tasks, recovering from errors, and assembling resources) are made according to local circumstances. However, making use of this approach would not be effective if it is not complemented with real time monitoring and control where timely manufacturing status can be used whenever available. Given such effective monitoring and controlling systems, the effectiveness of the planner to perform scheduling tasks will be much enhanced to the extent that real time reactive schedule-repair can be considered. In that regard, this scheduling strategy is conceptualized as Predictive-Reactive Scheduling. In analytical terms, the basic principles involved are:

- planning is done in advance using the most of the available information,

- reacting to earlier actions, as a consequence of earlier prediction, and considering new information as it become available.

These two steps are done iteratively and require heuristic, experience, intuition, and a lot of decision aids. The knowledge intensive work constitutes a domain in which application of knowledge-based techniques would be appealing.

The results in the implementation of a Predictive and Reactive Scheduling Tool Kit indicate that the strength of this tool kit is particular to the provision of a common window which shows the real-time information in the production office and enables the production staff to:

- visualize both the order status and material status and hence assess the impact of the alternative order release pattern in order to set the priority of order release

- keep an active account of inventory so as to limit and balance the stock level

- detect the possible material bottleneck and carry out relevant short term preventive action in order to streamline the material flow within the factory.

Empirical research results found that local autonomy, particularly in performing reactive scheduling, could be motivated by the provision of suitable decision aids and channels to access timely information. The Decision Support workbench prototype was targeted to establish a software environment (software process, knowledge, and tools, etc.) for user-oriented design and development of integrated PPC prototypes in order to react to earlier action autonomously. The three tools mentioned in this paper, namely Cumulative In / Out modelling, recursive BOM querying techniques and particularly, Petri-Net modelling were field tested to be the basic building blocks of the DS workbench. Our further research will focus on extracting knowledge in operational procedures and establishing process parameters in reactive scheduling. Object-Oriented Petri-Net Modelling tools is good at representing the manufacturing system will be developed to be one of the major knowledge acquisition elements in the workbench. Provided that the procedural functions and the related parameters are established first, knowledge-based techniques could be used more effectively under an evolving manufacturing system.

Acknowledgement:

The authors would like to acknowledge Wing Sang Bakelite Elec. Mfy. Co. Ltd. for the support of the case research settings in their company and express their thanks to the management team for their advices, insights and eager participation in the design and development of information systems.

References:

1. Thompshon, K., MRP II in Repetitive Manufacturing Environment, Production and Inventory Management, 4th Quarterly, pp.1-14, 1983.

2. Sepehri, M.P.E., and Raffish, N., Developing and implementing Control Systems for Repetitive Manufacturing, Industrial Engineering, June 1986.

3. Wemmerlöv, U., Assemble-to-Order manufacturing: Implication of materials Management, J. of Operations Management, Vol. 4, No.4, pp.347-368, 1984.

4. Bertrand, J.W.M., and Wijngaard, J., The Structuring of Production Control Systems, IJOPM 6(2), pp.5-20, 1986.

5. Hadavi, K.C., Shahraray, M.S., and Voigt, K., ReDS - A Dynamic Planning, Scheduling, and Control System for Manufacturing, J. of Mfg. Sys. 9(4), pp.332-344.

6. Hadavi, K.C., Autonomous Planning Agents for Real-time Control, in Recent Developments in Decision Support Systems, Clyde W. Holsapple and Andrew B. Whinston (eds.), Springer-Verlag, pp.521-533, 1993.

7. Sartori, L.G., Manufacturing Information Systems, Addison-Wesley Publishing Co., pp. 159-196, 1987.

8. Chan, M.K., Cheung, T.M., To, C.O., Tse, W.Y. and Wu, F., Procedural Knowledge Acquisition Using PetriNet Modelling Expert System Application, Higher Diploma (Mathematics, Statistics and Computing) advanced programming project report, Department of Applied Mathematics, Hong Kong Polytechnic, 1993.

9. Mather, H., Bills of Materials, APICS, 1987

Knowledge-Based Reactive Scheduling (B-15)
E. Szelke and R.M. Kerr (Editors)
Elsevier Science B.V. (North-Holland)

Open Discussion—A Summary

R. M. Kerr
Workshop Chairman

E. Szelke
Workshop Coordinator

The open discussion that followed the presentations focused on a variety of issues, starting with a stress on the importance of the complementary relationship between predictive and reactive scheduling. In connection with this, a particular area requiring further investigation was identified as being the nature of the predictive scheduling task (e.g., in terms of aggregation level) that should occur in a decentralized scheduling environment in which local reactive decisions are made by pseudo-autonomous scheduling agents.

The use of hybrid solution techniques in scheduling in general and reactive scheduling in particular was seen as being a potentially promising approach with different aspects of the scheduling problem being tackled using different problem representation and solution methods. Possible scope was seen for the use of evidential reasoning for combining different recommendations from different paradigms. It appeared to be also generally agreed that in the majority of present day manufacturing situations, the presence of the human in the problem solving loop and hence the effective integration of human and automated problem solving paradigms in some sort of cooperative mode was very important, possibly even more so for reactive than predictive scheduling. The human should always have the final decision in accepting or rejecting a schedule, and should often also play a dominant role in its construction or modification. This implies putting a considerable amount of effort into the development of user interfaces and of scheduling/rescheduling decision support systems rather than fully automated systems. A higher proportion of scheduling research should therefore be directed towards this area than is currently the case. It was proposed that the scheduling process involved a triangle of "human", "models" and "methods" with the human playing the dominant role, and that more attention should be placed on feedback from applications and implementation experience to develop explicit process models of scheduling support activities.

The subject of generalizability and re-usability of problem specific approaches was discussed, with the proposed use of the object oriented paradigm to develop a general purpose set of primitives that could be specialized to individual applications. Other proposals were for the development of domain-specific shells for particular manufacturing environments and the use of conceptual graphs in creating logic bases.

It was suggested during the course of the discussion that few of the reported reactive scheduling approaches gave much weight to technological constraints which tend to be a predominant feature of many factory environments. Also, a focus on accepting the reactive scheduling problem as given may be to focus on the wrong problem. With more attention to product and process rationalization, simplification, preventative maintenance etc., the incidence of reactive scheduling problems can be reduced and solutions made easier to find.

IFIP

The INTERNATIONAL FEDERATION FOR INFORMATION PROCESSING is a multinational federation of professional and technical organisations (or national groupings of such organisations) concerned with information processing. From any one country, only one such organisation – which must be representative of the national activities in the field of information processing – can be admitted as a Full Member. In addition a regional group of developing countries can be admitted as a Full Member. On 1 October 1993, 44 organisations were Full Members of the Federation, representing 65 countries.

The aims of IFIP are to promote information science and technology by:

- fostering international co-operation in the field of information processing;
- stimulating research, development and the application of information processing in science and human activity;
- furthering the dissemination and exchange of information about the subject;
- encouraging education in information processing.

IFIP is dedicated to improving worldwide communication and increased understanding among practitioners of all nations about the role information processing can play in all walks of life.

Information technology is a potent instrument in today's world, affecting people in everything from their education and work to their leisure and in their homes. It is a powerful tool in science and engineering, in commerce and industry, in education and adminstration. It is truly international in its scope and offers a significant opportunity for developing countries. IFIP helps to bring together workers at the leading edge of the technology to share their knowledge and experience, and acts as a catalyst to advance the state of the art.

IFIP came into official existence in January, 1960. It was established to meet a need identified at the first International Conference on Information Processing which was held in Paris in June, 1959, under the sponsorship of UNESCO.

Organisational Structure
The Federation is governed by a GENERAL ASSEMBLY, which meets once every year and consists of one representative from each Member organisation. The General Assembly decides on all important matters, such as general policy, the programme of activities, admissions, elections and budget.

IFIP ORGANISATION CHART

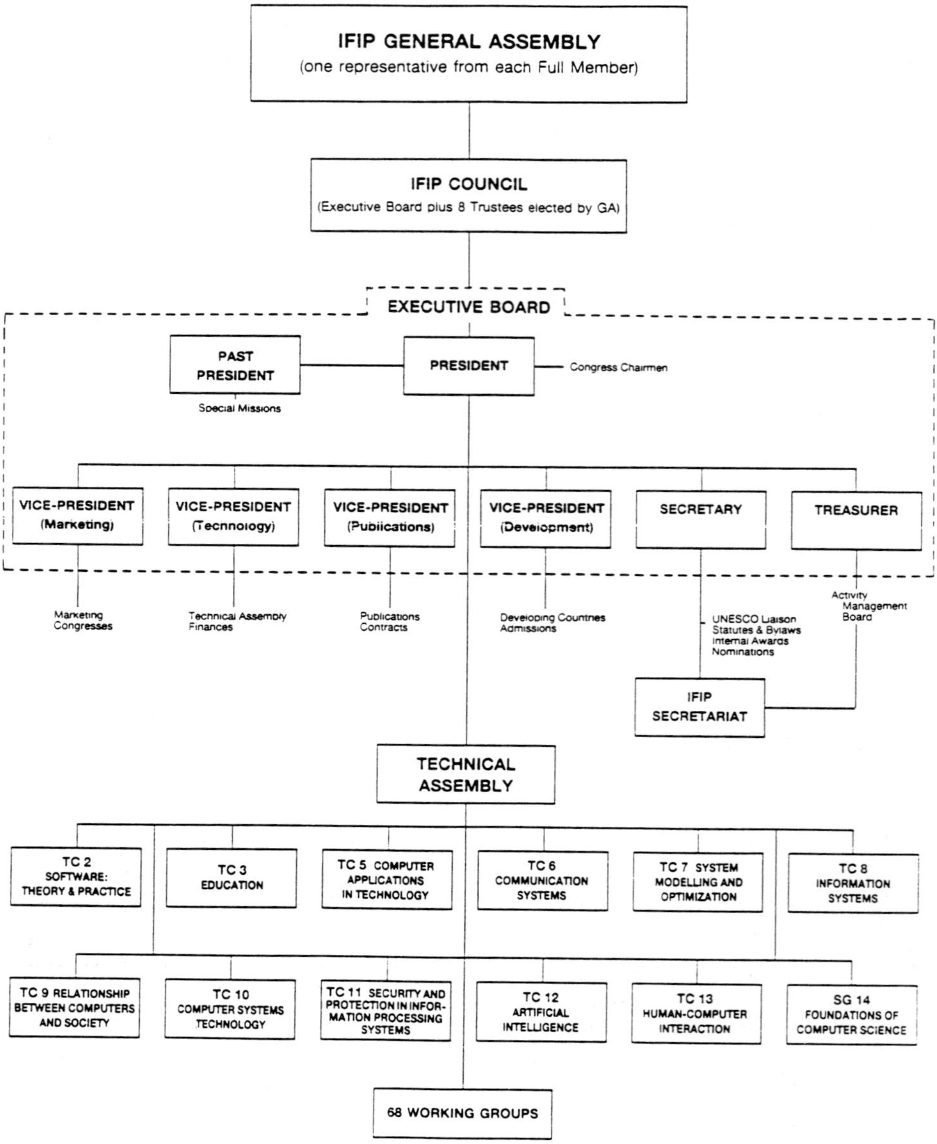

The day-to-day work of IFIP is directed by its Officers: the President, Vice-Presidents, Secretary and Treasurer, who are elected by the General Assembly and together constitute the EXECUTIVE BOARD.

The COUNCIL, consisting of the Officers and up to eight Trustees elected from the General Assembly, meets twice a year and takes decisions which become necessary between General Assembly meetings.

The headquarters of the Federation are in Geneva, Switzerland where the IFIP Secretariat administers its affairs.

For further information please contact:

IFIP Secretariat
attn. Mme. GWYNETH ROBERTS
16 Place Longemalle
CH-1204 Geneva, Switzerland
telephone: 41 (22) 310 26 49
facsimile: 41 (22) 781 23 22
Bitnet: ifip@cgeuge51

IFIP's MISSION STATEMENT

IFIP's mission is to be the leading, truly international, apolitical organisation which encourages and assists in the development, exploitation and application of Information Technology for the benefit of all people.

Principal Elements

1. To stimulate, encourage and participate in research, development and application of Information Technology (IT) and to foster international co-operation in these activities.
2. To provide a meeting place where national IT Societies can discuss and plan courses of action on issues in our field which are of international significance and thereby to forge increasingly strong links between them and with IFIP.
3. To promote international co-operation directly and through national IT Societies in a free environment between individuals, national and international governmental bodies and kindred scientific and professional organisations.
4. To pay special attention to the needs of developing countries and to assist them in appropriate ways to secure the optimum benefit from the application of IT.
5. To promote professionalism, incorporating high standards of ethics and conduct, among all IT practitioners.
6. To provide a forum for assessing the social consequences of IT applications; to campaign for the safe and beneficial development and use of IT and the protection of people from abuse through its improper application.

7. To foster and facilitate co-operation between academics, the IT industry and governmental bodies and to seek to represent the interest of users.
8. To provide a vehicle for work on the international aspects of IT development and application including the necessary preparatory work for the generation of international standards.
9. To contribute to the formulation of the education and training needed by IT practitioners, users and the public at large.

IFIP TRANSACTIONS

IFIP TRANSACTIONS is a serial consisting of 15,000 pages of valuable scientific information from leading researchers, published in 36 volumes per year. The serial includes contributed volumes, proceedings of the IFIP World Conferences, and conferences at Technical Committee and Working Group level. Mainstream areas in the IFIP TRANSACTIONS can be found in Computer Science and Technology, Computer Applications in Technology, and Communication Systems.

From 1993 onwards the IFIP TRANSACTIONS are only available as a full set.

IFIP TRANSACTIONS A:
Computer Science and Technology
1992: Volumes A1–A19
1993: Volumes A20–A40
1994: Volumes A41–A61
ISSN 0926-5473

IFIP Technical Committees that are involved in IFIP TRANSACTIONS A
Software: Theory and Practice (TC2)
Education (TC3)
System Modelling and Optimization (TC7)
Information Systems (TC8)
Relationship Between Computers and Society (TC9)
Computer Systems Technology (TC10)
Security and Protection in Information Processing Systems (TC11)
Artificial Intelligence (TC12)
Human-Computer Interaction (TC13)
Foundations of Computer Science (SG14)

IFIP TRANSACTIONS B:
Applications in Technology
1992: Volumes B1–B8
1993: Volumes B9–B14
1994: Volumes B15–B19
ISSN 0926-5481

IFIP Technical Committee that is involved in IFIP TRANSACTIONS B
Computer Applications in Technology (TC5)

IFIP TRANSACTIONS C:
Communication Systems
1992: Volumes C1–C8
1993: Volumes C9–C16
1994: Volumes C17–C26
ISSN 0926-549X

IFIP Technical Committee that is involved in IFIP TRANSACTIONS C
Communication Systems (TC6)

IFIP TRANSACTIONS FULL SET: A, B & C
1992: 35 Volumes. US $ 1892.00/Dfl. 3500.00
1993: 35 Volumes. US $ 2100.00/Dfl. 3885.00
1994: 36 Volumes. US $ 2277.00/Dfl. 4212.00
The Dutch Guilder prices (Dfl.) are definitive. The US $ prices mentioned above are for your guidance only and are subject to exchange rate fluctuations. Prices include postage and handling charges.

The volumes are also available separately in book form.

Please address all orders and correspondence to:

ELSEVIER SCIENCE B.V.
attn. PETRA VAN DER MEER
P.O. Box 103, 1000 AC Amsterdam
The Netherlands
telephone: 31 (20) 5862602
facsimile: 31 (20) 5862616
email: p.meer@elsevier.nl